Scientific Computation

Springer
Berlin
Heidelberg
New York
Hong Kong
London
Milan
Paris
Tokyo

Physics and Astronomy ONLINE LIBRARY

springeronline.com

Jean-Jacques Chattot

Computational Aerodynamics and Fluid Dynamics

An Introduction

With 80 Figures

 Springer

Professor Jean-Jacques Chattot

University of California
Department of Mechanical
and Aeronautical Engineering
One Shields Avenue
Davis, CA 95616, USA

Library of Congress Cataloging-in-Publication Data.
Chattot, J. J. Computational aerodynamics and fluid dynamics : an introduction/ Jean-Jacques Chattot. p. cm. –
(Scientific computation, ISSN 1434-8322) Includes bibliographical references and index.
1. Fluid dynamics – Data processing. 2. Fluid dynamics – Mathematical models. I. Title. II. Series.
QA911.C435 2002   532'.05 – dc21   2002021726

1st Edition 2002
Corrected 2nd Printing 2004

ISSN 1434-8322
ISBN 978-3-642-07798-2

Springer-Verlag is a part of Springer Science+Business Media

springeronline.com

© Springer-Verlag Berlin Heidelberg 2010
Printed in Germany

Cover design: *design & production* GmbH, Heidelberg

To my wife Adrienne
and son Eric

Preface

The field of computational fluid dynamics (CFD) has matured since the author was first introduced to electronic computation in the mid-sixties. The progress of numerical methods has paralleled that of computer technology and software. Simulations are used routinely in all branches of engineering as a very powerful means for understanding complex systems and, ultimately, improve their design for better efficiency.

Today's engineers must be capable of using the large simulation codes available in industry, and apply them to their specific problem by implementing new boundary conditions or modifying existing ones.

The objective of this book is to give the reader the basis for understanding the way numerical schemes achieve accurate and stable simulations of physical phenomena, governed by equations that are related, yet simpler, than the equations they need to solve. The model problems presented here are linear, in most cases, and represent the propagation of waves in a medium, the diffusion of heat in a slab, and the equilibrium of a membrane under distributed loads. Yet, regardless of the origin of the problem, the partial differential equations (PDE's) reflect the physical phenomena to be modeled and can be classified as being of hyperbolic, parabolic or elliptic type. The numerical treatment depends on the equation type that can represent several physical situations as diverse as heat conduction and viscous fluid flow. Nonlinear model problems are also presented and solved, such as the transonic small disturbance equation and the equations of gas dynamics. The model problems are given a full treatment, from the exact analytical solution, the analysis of the scheme's consistency and accuracy, the study of stability, to the detailed implementation of the scheme and of the boundary and/or initial conditions. It is the author's hope that this will entice the reader to write his/her own programs, and by doing so, learn more about CFD than a book can teach.

Davis, March 2002   *Jean-Jacques Chattot*

Table of Contents

1. Introduction

1.1 Motivation

The material in this book is based on lecture notes on computational fluid dynamics (CFD) that the author has developed over the past twenty years in France, at Centre National d'Etudes Supérieures de Mécanique and at the Université de Paris-Sud, and in the US at the University of California, Davis.

It is intended for senior undergraduate and first year graduate students who will be developing or using codes in the simulation of fluid flows or other physical phenomena governed by partial differential equations (PDEs).

It is the belief of the author, that a numerical method is not fully understood until it has been coded by the user and applied in simulation; each model and scheme in this book is presented with this goal in mind.

1.2 Content

The book is self contained and kept at a simple enough level that the reader will not need further references in order to understand the material.

The approach is based on the finite difference method (FD), which is widely employed as a method of discretization on cartesian mesh systems, in the physical domain, or in the computational domain after coordinate transformation. The extension to the finite volume method on arbitrary mesh sytems, including unstructured meshes, although feasible with a similar approach, would require all analyses to be performed numerically, instead of analytically in closed form, as is the case here.

The book is organized in chapters that build up each on material covered in the previous chapters, particularly Chaps. 2, 3 and 4 and Chaps. 8–11. Chapters 5, 6 and 7 can be read in any order.

The basics of the finite difference method are presented in Chap. 2. The tools that will be used throughout the book are introduced: the Taylor expansion and the complex mode analysis, which requires some complex algebra. They are the tools for the accuracy and stability analyses.

Chapter 3 is devoted to ordinary differential equations (ODEs) and their integration. ODEs represent an important particular case of partial differential equations (PDEs), when the number of independent variables reduces to one, either due to the nature of the physical problem (e.g. a time-dependent problem reaches a steady-state), or because the solution is expanded in terms of polynomials with unknown coefficients (Fourier series, etc...) in all but one independent variable.

Chapter 4 is a simple, but general discussion of PDEs, their type and classification, the notion of characteristic surfaces, compatibility relations and the jump conditions associated with conservation laws. For further understanding of this complex topic, a reference is recommended to the reader. This chapter is pivotal in making the connection between the physical phenomena of wave propagation, diffusion and equilibrium, and their mathematical counterparts, via the existence or non-existence of characteristics and their interpretation.

Chapters 5, 6 and 7 concern the linear model equations of hyperbolic, parabolic and elliptic type, respectively. Classical schemes are briefly reviewed and discussed in terms of accuracy and stability. Practical aspects of the implementation of selected schemes are presented; these should help the reader develop his/her own programs for the proposed methods. These models offer the opportunity to touch upon the subject of the solution of large linear algebraic systems of equations when implicit schemes and iterative techniques are discussed. The Thomas algorithm for tridiagonal matrices is described in detail.

Chapter 8 is devoted to the convection-diffusion equation, a model for the Navier-Stokes equations.

Chapter 9 presents the method of Murman and Cole, which, in many respects, is a precursor to the advanced CFD schemes of today. The author holds it in particular affection as he was asked as first assignment to implement it to simulate transonic flows past profiles and bodies of revolution at ONERA, France, in the early seventies.

Chapter 10, "treatment of nonlinearities" is a short complement in techniques of linearization, in an attempt to understand how linear stability analysis can still be of use in the context of nonlinear problems.

Chapter 11 represents the application of the previous material to a system, namely, the equations of gas dynamics, to study its type, its jumps and its solutions. An extension of the Murman-Cole scheme is introduced and results for a shock tube problem, the converging-diverging nozzle flow and the start-up of a supersonic wind tunnel, are given. This chapter contains advanced material such as eigenvalues and eigenfunctions. It can be skipped at the undergraduate level.

In each chapter, short problems are proposed to the reader, to illustrate a point or complete a proof that is only sketched. These problems should not require more than one page of derivations and calculations to complete. In Appendix A, more elaborate problems are proposed, taken from final exams in class. These should be completed in a two-hours time frame. The solutions to these problems are found in Appendix B.

Acknowledgement

The author is indebted to his first "Maîtres", Maurice Holt, Professor Emeritus, University of California Berkeley, USA, and Roger Peyret, Professor Emeritus, Universite de Nice, France, for their enthusiasm for CFD and their mentorship, that had such profound impact on his career.

2. Basics of the Finite-Difference Method

2.1 Representation of a Function by Discrete Values

Let $u = f(x)$ be a continuous and differentiable function to a sufficient order in the interval $[a, b]$. A discretization of $[a, b]$ with constant step h is introduced:

$$x_i = a + (i - 1)h, \quad x_1 = a, \quad x_{ix} = b, \quad h = \frac{b - a}{ix - 1},$$

ix is the total number of points of the discretization, h is the discretization step. Let $u_i = f(x_i)$ (see Fig. 2.1).

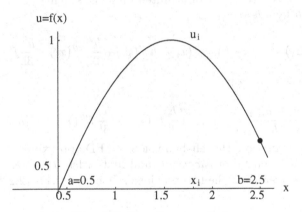

Fig. 2.1. Function $u = f(x)$

Remark: When $f(x)$ is given, the discrete values $u_i = f(x_i)$ are uniquely defined. However, if the discrete values $f(x_i)$ are given, it is not possible to find a unique $u = f(x)$ such that $u_i = f(x_i)$ without any further information about f (such as being a polynomial of some sort). Going from the continuous to the discrete carries with it a loss of information. In the finite difference method there is no hypothesis concerning the variation of the function between the points. This is in contrast to the finite-volume or finite-element methods where some assumption is made concerning the variation of the function between the points.

2.2 Representation of a First Derivative

Let $f(x)$ be a continuous function, differentiable as many times as needed in $[a, b]$. Then there exists a Taylor expansion about any point x_i:

$$u_{i+1} = f(x_{i+1}) = f(x_i + h) = f(x_i) + hf'(x_i) + \frac{h^2}{2!}f''(x_i) + \frac{h^3}{3!}f'''(x_i) + O(h^4).$$

$O(h^4)$ is the remainder and indicates that the unwritten terms are of fourth-order and higher. Since $f(x_i) = u_i$ we can write

$$\frac{u_{i+1} - u_i}{h} = f'(x_i) + \frac{h}{2!}f''(x_i) + \frac{h^2}{3!}f'''(x_i) + O(h^3). \tag{2.1}$$

As $h \to 0$, $(u_{i+1} - u_i)/h \to f'(x_i)$ thus the left-hand side of (2.1) is by definition a *finite-difference approximation* (FD) of the first derivative f' at point x_i. The leading term in the error is

$$\frac{h}{2!}f''(x_i) = O(h), \text{ as } h \to 0.$$

It is said that the scheme is first-order accurate. It is also qualified as a *one-sided* or *advanced* or *forward* finite-difference scheme.

Replace h by $-h$ and get

$$u_{i-1} = f(x_{i-1}) = f(x_i - h) = f(x_i) - hf'(x_i) + \frac{h^2}{2!}f''(x_i) - \frac{h^3}{3!}f'''(x_i) + O(h^4)$$

and

$$\frac{u_i - u_{i-1}}{h} = f'(x_i) - \frac{h}{2!}f''(x_i) + \frac{h^2}{3!}f'''(x_i) + O(h^3). \tag{2.2}$$

Again, by definition, the left-hand side is a FD approximation of f' at x_i. It is *one-sided* (*retarded* or *backward*) and first-order accurate.

Define the average of the two previous schemes: $((2.1) + (2.2))/2$

$$\frac{u_{i+1} - u_{i-1}}{2h} = f'(x_i) + \frac{h^2}{3!}f'''(x_i) + O(h^4). \tag{2.3}$$

This is a *centered* FD approximation for f' at x_i; it is second-order accurate. Note that the remainder is of fourth-order. This is because we have taken into account the fact that the odd-order derivatives vanish in the combination.

Non-centered schemes are not necessarily first-order accurate as the previous results may suggest. The following non-centered scheme is second-order accurate:

$$\frac{-u_{i+2} + 4u_{i+1} - 3u_i}{2h} = f'(x_i) - \frac{h^2}{3}f'''(x_i) + O(h^3).$$

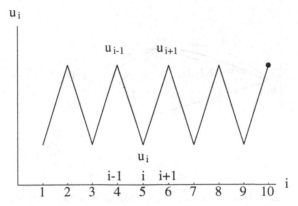

Fig. 2.2. Decoupling of odd and even points for the centered scheme with $f'(x_i) = 0$

2.3 Representation of a Second Derivative

Form the combination $((2.1)-(2.2))/h$:

$$\frac{u_{i+1} - 2u_i - u_{i-1}}{h^2} = f''(x_i) + \frac{h^2}{12} f^{(iv)}(x_i) + O(h^4). \qquad (2.4)$$

This is a centered and second-order accurate scheme for the second derivative f'' at point x_i. This scheme has a dominant coefficient for u_i, which is
an important feature when solving, for example, Laplace's equation, as will
be seen in Chap. 7.

2.4 Geometric Interpretation

The various schemes are displayed in Fig. 2.3. A parabola is fitted to the
three data points. Its slope is given exactly by scheme (2.3) and its second
derivative is given exactly by scheme (2.4).

Remarks: In the centered approximation (2.3), the value u_i does not appear. This can be a problem in certain cases and can lead to spurious solutions, such as the decoupling of odd and even points. In Fig. 2.2 the centered
scheme (2.3) yields $f'(x_i) = 0$, $i \geq 2$, even though the data are not constant.
There are two solutions, one going through the odd points and one going
through the even points, that satisfy $f'(x) = 0$ but with a different value for
the constant in $f(x) = \text{const}$. Schemes (2.1) and (2.2) "see" the decoupling
and do not allow it. There are several reasons for such decoupling to occur. It
can be triggered by large gradients in the solution, such as shocks, or due to
nonlinear effects. Loss of accuracy can also be a source for such decoupling,
as is the case with the solution to the singular boundary-value problem

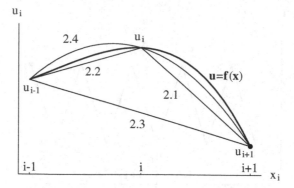

Fig. 2.3. Geometric representation of schemes (2.1)–(2.4)

$$\begin{cases} \epsilon\dfrac{d^2u}{dx^2} + \dfrac{du}{dx} = 2\epsilon + 2x, \\ \qquad\qquad u(0) = 0 \\ \qquad\qquad u(1) = 1. \end{cases} \qquad (2.5)$$

The exact solution is $u(x) = x^2$, regardless of the value ϵ. Using second-order accurate centered schemes for the derivatives, schemes (2.4) and (2.3), the exact solution satisfies the FDE identically as can be easily verified, by letting $u_i = x_i^2$ in

$$\epsilon\frac{u_{i+1} - 2u_i + u_{i-1}}{\Delta x^2} + \frac{u_{i+1} - u_{i-1}}{2\Delta x} - 2\epsilon - 2x_i = 0. \qquad (2.6)$$

However, when it comes to solving the FDE using a tridiagonal solver (see Chap. 6), for small enough values of ϵ the equation reduces to a first-order ODE with a centered scheme, prone to oscillations due to loss of accuracy. The result in Fig. 2.4 corresponds to $\epsilon = 10^{-9}$ and $\Delta x = 0.02$.

Problem: Perform the Taylor expansion of the FD quotient

$$\frac{u_{i+2} - u_i}{2h} = \dots, \qquad (2.7)$$

using the result of scheme (2.1). Then find the Taylor expansion of $\alpha(2.1) + \beta(2.7)$. Be sure to carry the expansions to $O(h^3)$. Find the coefficients α and β that will make the scheme second-order accurate for f_i'.

2.5 Taylor Expansion

The Taylor expansion in two or more variables can be obtained from the Taylor expansion in one variable, by repeating the operation, variable by variable. One obtains for two variables:

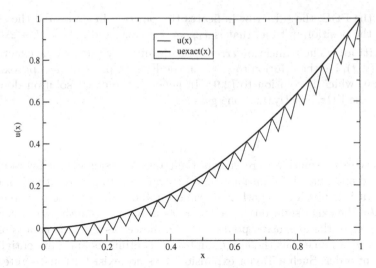

Fig. 2.4. Odd-even decoupling due to loss of accuracy

$$f(x_{i+1}, y_{j+1}) = f_{i,j} + \Delta x \frac{\partial f_{i,j}}{\partial x} + \Delta y \frac{\partial f_{i,j}}{\partial y} + \frac{\Delta x^2}{2} \frac{\partial^2 f_{i,j}}{\partial x^2} + \Delta x \Delta y \frac{\partial^2 f_{i,j}}{\partial x \partial y}$$

$$+ \frac{\Delta y^2}{2} \frac{\partial^2 f_{i,j}}{\partial y^2} + \frac{\Delta x^3}{3!} \frac{\partial^3 f_{i,j}}{\partial x^3} + \frac{\Delta x^2 \Delta y}{2!} \frac{\partial^3 f_{i,j}}{\partial x^2 \partial y}$$

$$+ \frac{\Delta x \Delta y^2}{2!} \frac{\partial^3 f_{i,j}}{\partial x \partial y^2} + \frac{\Delta y^3}{3!} \frac{\partial^3 f_{i,j}}{\partial y^3} + O(\Delta x + \Delta y)^4.$$

Using the binomial formula this can be written as

$$f_{i+1,j+1} = f_{i,j} + \dots + \frac{1}{n!} \left(\Delta x \frac{\partial}{\partial x} + \Delta y \frac{\partial}{\partial y} \right)^n f_{i,j} + \dots .$$

2.6 Consistency and Accuracy

We introduce the notion of *truncation error* (TE). This is an important element in the qualification of a finite-difference scheme. We illustrate this with an example. Consider the partial differential equation (PDE), the heat equation:

$$\frac{\partial u}{\partial t} = \alpha \frac{\partial^2 u}{\partial x^2}, \ \alpha > 0, \tag{2.8}$$

and the following finite-difference equation (FDE) written for a zero right-hand side as

$$\frac{u_i^{n+1} - u_i^n}{\Delta t} - \alpha \frac{u_{i+1}^n - 2u_i^n + u_{i-1}^n}{\Delta x^2} = 0. \tag{2.9}$$

In the FDE, the subscript indicates the position in space and the superscript the position in time, that is $x_{i+1} = x_i + \Delta x$, and $t^{n+1} = t^n + \Delta t$.

Definition: The truncation error ϵ_i^n is obtained by putting the exact solution $\widehat{u}(x,t)$, i.e. the solution to (2.8), in the FDE in place of the approximate solution, which is solution to (2.9). In general, the exact solution does not satisfy the FDE exactly, and one gets

$$\epsilon = \frac{\widehat{u}_i^{n+1} - \widehat{u}_i^n}{\Delta t} - \alpha \frac{\widehat{u}_{i+1}^n - 2\widehat{u}_i^n + \widehat{u}_{i-1}^n}{\Delta x^2} \neq 0.$$

If the exact solution were known, the true expression of the TE would be obtained and much information would be available. In general, this is not the case, but it is possible to gather information about the scheme by expanding *formally* the exact solution in Taylor series about the point (i,n), indicating the choice of the expansion point with the indices as ϵ_i^n. This is possible, so long as we assume that the exact solution is continuous and differentiable to sufficient order. Such a Taylor expansion does not exist for the discrete solution u_i^n. Here we obtain (using (2.1) and (2.4) with the appropriate changes of variables)

$$\epsilon_i^n = \left(\frac{\partial \widehat{u}_i^n}{\partial t} - \alpha \frac{\partial^2 \widehat{u}_i^n}{\partial x^2} \right) + \frac{\Delta t}{2} \frac{\partial^2 \widehat{u}_i^n}{\partial t^2} - \alpha \frac{\Delta x^2}{12} \frac{\partial^4 \widehat{u}_i^n}{\partial x^4} + O(\Delta t^2, \Delta x^4).$$

Since the exact solution satisfies the PDE, the terms in the first bracket cancel and the TE reduces to

$$\epsilon_i^n = \frac{\Delta t}{2} \frac{\partial^2 \widehat{u}_i^n}{\partial t^2} - \alpha \frac{\Delta x^2}{12} \frac{\partial^4 \widehat{u}_i^n}{\partial x^4} + O(\Delta t^2, \Delta x^4).$$

To be able to conclude the analysis, the expansion must be carried far enough to include the leading terms. Here, the leading terms in the TE are of order $O(\Delta t, \Delta x^2)$, which means that they dominate the error ϵ_i^n for small values of Δt and Δx.

Definition: A finite-difference scheme is said to be *consistent* if $\epsilon_i^n \to 0$ as Δt, $\Delta x \to 0$ independently. If the truncation error is of the form $\epsilon_i^n = O(\Delta t^p, \Delta x^q)$, p, $q > 0$ the scheme is said to be of order p in t and of order q in x.

Here $p = 1$, $q = 2$. The scheme is first-order accurate in time and second-order accurate in space.

Remark: The notion of consistency and accuracy is independent of the point chosen for the Taylor expansion. Indeed, as mentioned above, ϵ is obtained by introducing the exact solution in the FDE. Expanding the exact solution about any point in the vicinity of (i,n) will produce the same value for ϵ, that is if the full expansion was carried out. Since we limit ourselves to a few terms, it suffices that the leading terms in the TE be calculated. These terms will have the same asymptotic behavior, regardless of the point chosen for the expansion.

Remark: The TE must go to zero in the definition of a consistent scheme, regardless of how the discretization step goes to zero: this is the reason for the term *independently*. A famous counterexample is due to DuFort and Frankel:

$$\frac{u_i^{n+1} - u_i^{n-1}}{2\Delta t} = \alpha \frac{u_{i+1}^n - u_i^{n+1} - u_i^{n-1} + u_{i-1}^n}{\Delta x^2}. \tag{2.10}$$

Adding and subtracting $2u_i^n$, the TE reads (the 'hat' has been dropped for simplicity):

$$\epsilon_i^n = \frac{u_i^{n+1} - u_i^{n-1}}{2\Delta t} - \alpha \left(\frac{u_{i+1}^n - 2u_i^n + u_{i-1}^n}{\Delta x^2} - \frac{\Delta t^2}{\Delta x^2} \frac{u_i^{n+1} - 2u_i^n + u_i^{n-1}}{\Delta t^2} \right)$$

Now, each term is easily identifiable with a known FD approximation ((2.3) and (2.4)) for which the Taylor expansions have been performed about point (i, n) as

$$\epsilon_i^n = \frac{\partial u_i^n}{\partial t} + O(\Delta t^2) - \alpha \frac{\partial^2 u_i^n}{\partial x^2} + O(\Delta x^2) - \alpha \frac{\Delta t^2}{\Delta x^2} \frac{\partial^2 u_i^n}{\partial t^2} + O\left(\frac{\Delta t^4}{\Delta x^2} \right).$$

The first two partial derivatives cancel out since $u(x, t)$ is the exact solution to the PDE. The TE reduces to

$$\epsilon_i^n = -\alpha \frac{\Delta t^2}{\Delta x^2} \frac{\partial^2 u_i^n}{\partial t^2} + O(\Delta t^2) + O(\Delta x^2) + O\left(\frac{\Delta t^4}{\Delta x^2} \right).$$

If $\Delta t \to 0$ faster than Δx, that is if there exists $r > 0$ such that $\Delta t = O(\Delta x^{1+r})$, then $\epsilon_i^n = O(\Delta x^{2r}, \Delta x^2)$. The scheme is consistent and of order $\min(2r, 2)$.

If $\Delta t \to 0$ as Δx or slower, then the scheme is not consistent. For instance, if $\Delta t = K\Delta x$ then $\epsilon_i^n = -\alpha K^2 \frac{\partial^2 u_i^n}{\partial t^2} + O(\Delta x^2)$, where the leading term is in general different from zero, except at steady-state. The DuFort–Frankel scheme is not consistent with the heat equation, an equation of parabolic type, but rather with the following equation:

$$\frac{\partial u}{\partial t} + \alpha K^2 \frac{\partial^2 u}{\partial t^2} = \alpha \frac{\partial^2 u}{\partial x^2},$$

which is of hyperbolic type.

Problem: Recast the FD quotient and use previous Taylor expansions to find which derivative is approximated and to what order by

$$\frac{u_{i+1,j+1} - u_{i,j+1} - u_{i+1,j} + u_{i,j}}{\Delta x \Delta y}.$$

2.7 Stability

Consistency and stability are the two fundamental properties of a useful scheme, as stated in *Lax's Equivalence Theorem:*

Theorem: Given a properly posed initial value problem and a finite difference approximation to it that satisfies the consistency condition, stability is the necessary and sufficient condition for convergence.

This theorem has not been proved for general nonlinear equations.

Convergence means that, in a certain norm, the difference between the exact and the discrete solutions $\|\widehat{u}(x_i, t^n) - u_i^n\|$ will tend to zero as the small parameters (Δt, Δx, ...) tend to zero.

Stability is concerned with round-off errors occuring in the solution of the FDE on a computer with finite accuracy. When the FDE is a linear, non-homogeneous equation, the errors can be shown to satisfy the same linear equation, made homogeneous, for which the zero solution is expected to be a stable state (an attractor). As we have seen, the main tool for consistency/accuracy is the Taylor expansion. For stability we shall use the *Von Neumann analysis*. Other techniques exist, such as matrix eigenvalues analysis, but they are more involved. The main advantage of the Von Neumann method is its simplicity and flexibility. It is a local analysis of wave mode amplification of the round-off errors. It does not take into account the global problem with its boundary conditions.

Note that Von Neumann analysis can only be applied to a linear homogeneous equation or system, as it is based on superposition of wave modes. Nonlinear equations must first be linearized by "freezing" the coefficients of the partial derivatives.

Consider again the heat equation in *update* form (that is the form used to calculate the new values of the unknowns):

$$u_i^{n+1} = u_i^n + \alpha \frac{\Delta t}{\Delta x^2}(u_{i+1}^n - 2u_i^n + u_{i-1}^n).$$

The study is carried out on each Fourier mode separately and the linearity of the equation insures that, if the scheme is stable for each mode, it will be stable for any superposition of modes. Time and space are treated differently; introduce the complex Fourier mode:

$$u_i^n = g^n e^{\underline{i}i\beta} = g^n(\cos i\beta + \underline{i}\sin i\beta),$$

where g is the complex amplitude raised to the n-th power (n corresponds to time), β is the wave number corresponding to the Fourier mode (i corresponds to space) and $\underline{i}^2 = -1$. Putting the complex Fourier mode into the equation yields:

$$g^{n+1}e^{\underline{i}i\beta} = g^n e^{\underline{i}i\beta} + \sigma g^n(e^{\underline{i}(i+1)\beta} - 2e^{\underline{i}i\beta} + e^{\underline{i}(i-1)\beta}),$$

where $\sigma = \alpha(\Delta t)/(\Delta x^2)$. Dividing through by $g^n e^{\iota i \beta}$ gives the amplification factor

$$g = 1 - 2\sigma(1 - \cos\beta).$$

In order for the scheme to be stable, i.e. for the round-off errors to be damped, one must have $|g| \leq 1$, $\forall\beta$. In other words:

$$-1 \leq 1 - 2\sigma(1 - \cos\beta) \leq 1, \ \forall\beta.$$

The right inequality is always satisfied since $\sigma \geq 0$. The left inequality yields $\sigma(1 - \cos\beta) \leq 1$ which will be satisfied if $\sigma \leq \frac{1}{2}$. In terms of the time step this means:

$$\Delta t \leq \frac{\Delta x^2}{2\alpha}. \tag{2.11}$$

This is a restrictive condition on the time step, once the space step has been chosen. It is quite severe since, if the space step is divided by two, the maximum time step will be divided by four. There is an underlying physical and mathematical reason for this condition, that will become apparent when we study this parabolic equation in Chap. 6.

Consider now the linear convection equation

$$\frac{\partial u}{\partial t} + c\frac{\partial u}{\partial x} = 0. \tag{2.12}$$

Lax proposed, in 1954, the following scheme:

$$\frac{u_i^{n+1} - \frac{u_{i-1}^n + u_{i+1}^n}{2}}{\Delta t} + c\frac{u_{i+1}^n - u_{i-1}^n}{2\Delta x} = 0.$$

We now use the techniques presented above to study the consistency and stability of Lax' scheme.

Again, we can rewrite this equation differently to simplify the derivation of the TE:

$$\frac{u_i^{n+1} - u_i^n}{\Delta t} - \frac{u_{i+1}^n - 2u_i^n + u_{i-1}^n}{2\Delta t} + c\frac{u_{i+1}^n - u_{i-1}^n}{2\Delta x} = 0$$

where the central term has been obtained by adding and subtracting $2u_i^n$ to the numerator of the time derivative term. Using the previously derived Taylor expansions yields

$$\epsilon_i^n = \frac{\partial u_i^n}{\partial t} + \frac{\Delta t}{2}\frac{\partial^2 u_i^n}{\partial t^2} + O(\Delta t^2) - \frac{\Delta x^2}{2\Delta t}\frac{\partial^2 u_i^n}{\partial x^2} + O\left(\frac{\Delta x^4}{\Delta t}\right) + c\frac{\partial u_i^n}{\partial x} + O(\Delta x^2).$$

The first and last partial derivatives cancel to leave as leading terms

$$\epsilon_i^n = \frac{\Delta t}{2}\frac{\partial^2 u_i^n}{\partial t^2} - \frac{\Delta x^2}{2\Delta t}\frac{\partial^2 u_i^n}{\partial x^2} + O\left(\Delta x^2, \frac{\Delta x^4}{\Delta t}\right). \tag{2.13}$$

Consistency requires that $\Delta t = O(\Delta x^{2-r})$, $0 < r < 2$, in which case the TE reduces to $\epsilon_i^n = -(\Delta x^r)/2(\partial^2 u_i^n)/(\partial t^2) + O(\Delta x^{2-r})$. Lax' scheme is conditionally consistent. It is interesting to note that, unlike the DuFort and Frankel scheme, Lax' scheme requires large time steps for consistency. We expect, however, that stability will put a restriction on how large a time step we can choose.

Let $u_i^n = g^n e^{\underline{i} i \beta}$ and $\sigma = c(\Delta t)/(\Delta x)$. In update form the equation reads

$$u_i^{n+1} = \frac{1}{2}(u_{i-1}^n + u_{i+1}^n) - \frac{\sigma}{2}(u_{i+1}^n - u_{i-1}^n)$$

and the amplification factor is

$$g = \cos\beta - \underline{i}\sigma\sin\beta.$$

Taking the square of the modulus

$$|g|^2 = \cos^2\beta + \sigma^2\sin^2\beta = 1 - (1 - \sigma^2)\sin^2\beta.$$

The condition for stability reduces to $|\sigma| \le 1$. The limitation on the time step is

$$\Delta t \le \frac{\Delta x}{|c|}, \tag{2.14}$$

which is known as the *Courant-Friedrichs-Lewy condition* (CFL) (1928). An interpretation of this condition will be given in Chap. 5.

Problem: Study the consistency and accuracy of the following scheme for the linear convection equation

$$\frac{u_i^{n+1} - u_i^n}{\Delta t} + c\frac{-u_{i+1}^n + 4u_i^n - 3u_{i-1}^n}{2\Delta x} = 0,$$

and give the condition on $\sigma = c(\Delta t)/(\Delta x)$ for the scheme to be stable.

2.8 Complements on Truncation Error

It has been said that the truncation error is independent of the point chosen to evaluate it. We want to illustrate this point and at the same time give some shortcuts in obtaining the TE. It can be cumbersome and a source of errors or of unseen cancellations to perform directly the TE for each term about a single point, say (i, n). It is advised to proceed in the following way:

i) expand each finite-difference grouping representing one partial derivative with respect to the most convenient point for that term,

ii) then expand the resulting Taylor expansion with respect to the point chosen for the full FDE, by performing a shifting operation (i.e. a Taylor expansion) of each term, carrying the expansion far enough to retain the global order desired of the remainder.

Example: Consider the heat equation and the scheme introduced earlier (2.9). We have seen that

$$\epsilon_i^n = \frac{\Delta t}{2}\frac{\partial^2 u_i^n}{\partial t^2} - \alpha\frac{\Delta x^2}{12}\frac{\partial^4 u_i^n}{\partial x^4} + O(\Delta t^2, \Delta x^4) = O(\Delta t, \Delta x^2) \qquad (2.15)$$

We now choose to find the TE by expanding about point $(i, n+1)$. We proceed with each term separately. The first term is easily expanded about $(i, n+1)$ as a retarded FD to give

$$\frac{u_i^{n+1} - u_i^n}{\Delta t} = \frac{\partial u_i^{n+1}}{\partial t} - \frac{\Delta t}{2}\frac{\partial^2 u_i^{n+1}}{\partial t^2} + O(\Delta t^2).$$

The second term is best expanded about (i, n) using (2.4)

$$\frac{u_{i+1}^n - 2u_i^n + u_{i-1}^n}{\Delta x^2} = \frac{\partial^2 u_i^n}{\partial x^2} + \frac{\Delta x^2}{12}\frac{\partial^4 u_i^n}{\partial x^4} + O(\Delta x^4).$$

Now we shift the two terms from (i, n) to $(i, n+1)$ by noting that $n = (n+1)-1$, or in other words that $t^n = t^{n+1} - \Delta t$ and expanding with respect to $-\Delta t$ as

$$\frac{\partial^2 u_i^n}{\partial x^2} = \frac{\partial^2 u_i^{n+1}}{\partial x^2} - \Delta t\frac{\partial^3 u_i^{n+1}}{\partial t\partial x^2} + O(\Delta t^2)$$

$$\frac{\Delta x^2}{12}\frac{\partial^4 u_i^n}{\partial x^4} = \frac{\Delta x^2}{12}\left(\frac{\partial^4 u_i^{n+1}}{\partial x^4} - \Delta t\frac{\partial^5 u_i^{n+1}}{\partial t\partial x^4} + O(\Delta t^2)\right).$$

Note that the second term in the bracket is already of higher order than needed and will be put in the remainder. Combining these last results yields

$$\epsilon_i^{n+1} = \frac{\partial u_i^{n+1}}{\partial t} - \frac{\Delta t}{2}\frac{\partial^2 u_i^{n+1}}{\partial t^2} + O(\Delta t^2)$$
$$-\alpha\left(\frac{\partial^2 u_i^{n+1}}{\partial x^2} - \Delta t\frac{\partial^3 u_i^{n+1}}{\partial t\partial x^2} + O(\Delta t^2) + \frac{\Delta x^2}{12}\frac{\partial^4 u_i^{n+1}}{\partial x^4} + O(\Delta t\Delta x^2)\right).$$

Rearranging the result gives

$$\epsilon_i^{n+1} = -\frac{\Delta t}{2}\frac{\partial^2 u_i^{n+1}}{\partial t^2} + \alpha\Delta t\frac{\partial^3 u_i^{n+1}}{\partial t\partial x^2} - \alpha\frac{\Delta x^2}{12}\frac{\partial^4 u_i^{n+1}}{\partial x^4} + O(\Delta t^2, \Delta t\Delta x^2)$$
$$= O(\Delta t, \Delta x^2). \qquad (2.16)$$

Comparing (2.15) to (2.16) shows that indeed we obtain the same result for consistency and accuracy when expanding about (i, n) and $(i, n+1)$.

We can also ask ourselves what is the accuracy of a scheme for which the TE reads $\epsilon = O(\Delta x, \Delta t\Delta x, \Delta t^2)$? Such a scheme is clearly first-order in x, but is it first- or second-order in t? To illustrate this point, consider the linear convection equation (2.12) with $c = 1$, discretized with a "leap-frog" scheme in time and retarded scheme in space:

$$\frac{u_i^{n+1} - u_i^{n-1}}{2\Delta t} + \frac{u_i^n - u_{i-1}^n}{\Delta x} = 0. \tag{2.17}$$

The TE about point (i, n) is easily obtained from previous results as

$$\epsilon_i^n = \frac{\Delta t^2}{3!}\frac{\partial^3 u_i^n}{\partial t^3} - \frac{\Delta x}{2}\frac{\partial^2 u_i^n}{\partial x^2} + \frac{\Delta x^2}{3!}\frac{\partial^3 u_i^n}{\partial x^3} + O(\Delta x^3, \Delta t^4) = O(\Delta x, \Delta t^2). \tag{2.18}$$

The result indicates that this scheme is first-order accurate in x and second-order accurate in t.

Now, we evaluate the TE at $(i, n + 1)$. We perform the shifting of the Taylor expansion for each term in (2.18) to obtain

$$\epsilon_i^{n+1} = -\frac{\Delta x}{2}\frac{\partial^2 u_i^{n+1}}{\partial x^2} + \frac{\Delta t \Delta x}{2}\frac{\partial^3 u_i^{n+1}}{\partial t \partial x^2} + \frac{\Delta t^2}{3!}\frac{\partial^3 u_i^{n+1}}{\partial t^3}$$
$$+ \frac{\Delta x^2}{3!}\frac{\partial^3 u_i^{n+1}}{\partial x^3} + O(\Delta t + \Delta x)^3, \tag{2.19}$$

which has the form in question. The scheme is still first-order in space and second-order in time, even though it seems that, as $\Delta t \to 0$, Δx being held fixed, ϵ_i^{n+1} has a term proportional to Δt. To understand the mechanism, it suffices to study the TE, plugging an exact solution into the scheme. Let $u(x, t) = (x - t)^3$ be the exact solution to (2.17). By definition

$$\epsilon = \frac{(x_i - t^{n+1})^3 - (x_i - t^{n-1})^3}{2\Delta t} + \frac{(x_i - t^n)^3 - (x_{i-1} - t^n)^3}{\Delta x}.$$

After some algebra this can be written as

$$\epsilon = -3(x_i - t^n)\Delta x - \Delta t^2 + \Delta x^2$$

or as

$$\epsilon = -3(x_i - t^{n+1})\Delta x - 3\Delta t \Delta x - \Delta t^2 + \Delta x^2.$$

Note that the coefficients in the TE are the partial derivatives of the exact solution as obtained in (2.18) and (2.19). The choice of a low-order polynomial is the reason for having only a few terms in ϵ since all partial derivatives in the solution of fourth- or higher-order vanish.

To come back to the discussion, as seen above the error is unique, and the presence of the $\Delta t \Delta x$ term results from the choice of the expansion point. Another way of assessing the accuracy of the scheme in time is to have an infinitely accurate scheme in x by letting $\Delta x \to 0$ and study the behavior of the TE with respect to Δt. We find that

$$\epsilon = \frac{\Delta t^2}{3!}\frac{\partial^3 u_i^{n,n+1}}{\partial t^3} = -\Delta t^2,$$

which confirms that the scheme is second-order accurate in time.

By choosing the expansion point (typically a center of symmetry for the scheme), it is possible to get rid of the cross-derivative term (the second term in (2.19)).

The basic mathematical tools that we have used in this chapter and that we will be using in the rest of the book are the Taylor expansion and complex numbers algebra, the former being used for the study of consistency and accuracy, the latter for the analysis of stability. These notions are assumed to be known. They can be reviewed in various textbooks such as Reference [1].

3. Application to the Integration
of Ordinary Differential Equations

3.1 Introduction

Ordinary Differential Equations (ODE) play an important role in mathematics for the engineers. This is due to the fact that many laws of physics can be expressed as ODEs.

An ODE is a relation between one or several derivatives with respect to x of an unknown scalar function $u(x)$. This relation can also depend on u itself, on given functions of x and on constants. The independent variable, x, can represent space, time, or any physical variable.

Definition: An ordinary differential equation is said to be of order n if the highest derivative in the equation is $\frac{d^n u}{dx^n}$.

Any ODE of order n can be transformed into a system of n ODEs of first-order upon introducing $n-1$ new unknowns which are the successive derivatives of u, $v = u'$, $w = v'$, etc..., and the associated equations. The converse is not always possible.

A first-order ODE is of the general implicit form $F(x, u, u') = 0$.

In most cases, it can be solved for u' explicitly as

$$u' = f(x, u). \tag{3.1}$$

This relation expresses the slope of the integral curve $u(x)$ in the (x, u) plane. There exist several techniques for solving (3.1) analytically. Notions of existence and uniqueness, of general solution, and of particular solution can be found in [1]. We will assume that the problem is well posed for (3.1), when complemented by an *initial condition*

$$u(x_0) = u_0. \tag{3.2}$$

Problem (3.1)+(3.2) is called an *initial value problem*.

In general an ODE of order n requires n *initial conditions*.

Remarks: When x represents time or is time-like, in general one must solve an initial value problem.

When x is space-like, the boundary conditions can be split between both ends of the domain. Furthermore, a nonlinear first-order ODE may require more than one condition, i.e. more conditions than the order of the equation.

This is the case for boundary value problems of mixed-type for an equation or system of equations as we will see in Chaps. 8 to 11. At this point an example will suffice. The following ODE

$$\frac{d}{dx}\left(\frac{u^2}{2}\right) = u, \ 0 \le x \le 1, \tag{3.3}$$

with two boundary conditions

$$u(0) = 1, \ u(1) = -1, \tag{3.4}$$

has the exact solution

$$\begin{cases} u(x) = x + 1, \ 0 \le x < \frac{1}{2} \\ u(x) = x - 2, \ \frac{1}{2} < x \le 1. \end{cases}$$

The exact and numerical solutions are shown in Fig. 3.1. Note the jump at $x = \frac{1}{2}$. The numerical solution has been obtained using a mixed-type scheme of Chapter 10.

We shall now consider various numerical methods for solving (3.1)–(3.2).

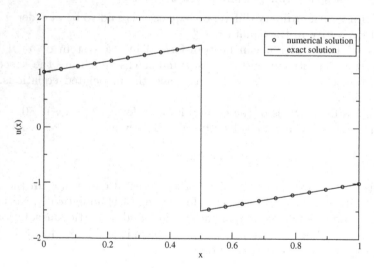

Fig. 3.1. Two-points boundary value problem for a nonlinear first-order ODE

3.2 The Euler–Cauchy Method

In this method, an advanced FD scheme is used for the derivative, and the right-hand side is evaluated at the initial values of the step. Let h be the constant integration step (discretization step), and $x_i = x_0 + (i - 1)h$. For the problem (3.1)–(3.2), the method reads:

$$\begin{cases} \dfrac{u_1 - u_0}{h} = f(x_0, u_0) \\ \qquad\qquad \dots \\ \dfrac{u_{i+1} - u_i}{h} = f(x_i, u_i). \end{cases} \qquad (3.5)$$

In update form the general formula is $u_{i+1} = u_i + hf(x_i, u_i)$, $i = 0, 1, \dots$

The simple interpretation of the scheme is that the "new" value u_{i+1} is obtained as the intersection of the tangent to the integral curve at point (x_i, u_i) and the coordinate line $x = x_{i+1}$. The analysis of the Euler-Cauchy method can be carried out with the truncation error, from the differential form (3.5) as

$$\epsilon_i = \frac{u_{i+1} - u_i}{h} - f(x_i, u_i) = u_i' + \frac{h}{2} u_i'' - f(x_i, u_i) + O(h^2).$$

The first and third terms cancel out since u is the exact solution to (3.1). The leading term in the TE is of order h, thus the Euler-Cauchy method is first-order accurate.

Note that, if the curvature of the integral curve is zero, i.e. the integral curve is a straight line, the error is zero and the exact solution is obtained at each integration point.

3.3 Improved Euler Method

This method aims at improving the accuracy of the previous method. It is a two-step method.

First a provisional value using the Euler-Cauchy method is computed as

$$\tilde{u}_{i+1} = u_i + hf(x_i, u_i).$$

Then, the final value is obtained

$$u_{i+1} = u_i + \frac{h}{2} \left(f(x_i, u_i) + f(x_{i+1}, \tilde{u}_{i+1}) \right).$$

Upon elimination of the provisional value $\tilde{u}_{i+1}$ from the scheme, the TE reads:

$$\epsilon_i = \frac{u_{i+1} - u_i}{h} - \frac{1}{2} \left(f(x_i, u_i) + f \left[x_i + h, u_i + hf(x_i, u_i) \right] \right).$$

We use the Taylor expansion formula for a function of two variables $f(x, u)$ to get

$$\epsilon_i = u_i' + \frac{h}{2} u_i'' + \frac{h^2}{3!} u_i''' + O(h^3) - f(x_i, u_i) - \frac{h}{2} \frac{\partial f_i}{\partial x} - \frac{h f_i}{2} \frac{\partial f_i}{\partial u} + O(h^2),$$

where $f_i = f(x_i, u_i)$.

Rearranging the terms yields

$$\epsilon_i = \frac{h}{2}\left(u_i'' - \frac{\partial f_i}{\partial x} - f_i\frac{\partial f_i}{\partial u}\right) + O(h^2).$$

The term in bracket is identically zero since it represents $(d/dx)(u' - f(x,u))$, where the chain rule is used to calculate the total derivative of f.

Note that we did not push the expansion far enough to collect all the terms of second-order. If we had, we would have concluded, since in general they do not cancel out, that the scheme is second-order accurate. From our result above we can say that the improved Euler method is at least second-order accurate.

Remark: The improved Euler method will integrate exactly a polynomial of order two.

Problem: Integrate by hand $u' = 2x$, $u(0) = 1$ for a few steps and show that the discrete solution coincides with the exact solution.

3.4 The Runge–Kutta Method

We will limit ourselves to the fourth-order Runge-Kutta method. It is very popular because of its high accuracy. It is the basis of many software packages which allow for adjustable integration steps with control of the integration error. With a fixed value of h, the four steps of the Runge-Kutta method are

$$\begin{cases} a_i = hf(x_i, u_i) \\ b_i = hf(x_i + \frac{h}{2}, u_i + \frac{a_i}{2}) \\ c_i = hf(x_i + \frac{h}{2}, u_i + \frac{b_i}{2}) \\ d_i = hf(x_i + h, u_i + c_i) \end{cases} \tag{3.6}$$

to give

$$u_{i+1} = u_i + \frac{1}{6}(a_i + 2b_i + 2c_i + d_i). \tag{3.7}$$

It can be shown that the Runge-Kutta method is fourth-order accurate.

Remark: If f does not depend on u, i.e. $\partial f/\partial u = 0$, (3.6) and (3.7) reduce to Simpson's rule of integration.

3.5 Integration of Polynomials

A method of order p for (3.1) integrates exactly polynomials of order up to p.

Assume that the exact solution to (3.1) is a polynomial of degree p, $u(x) = P_p(x)$. By definition

$$u'(x) = \frac{dP_p(x)}{dx} = f(x),$$

where the right-hand side is a polynomial of degree $p - 1$. An integration method of order-p satisfies

$$\epsilon_i = \frac{u_{i+1} - u_i}{h} - \left\{\sum f\right\} = O(h^p),$$

where the term in bracket represents the numerical approximation to the right-hand side of the differential equation. Multiplying by h gives

$$h\epsilon_i = u_{i+1} - u_i - h\left\{\sum f\right\} = O(h^{p+1}).$$

The expression $u_{i+1} - u_i$ can be replaced by its Taylor expansion

$$u_{i+1} - u_i = \sum_{n=1}^{p} \frac{h^n}{n!} u_i^{(n)}.$$

Hence, $u_{i+1} - u_i - h\left\{\sum f\right\}$ is at most of degree p in h and the equality $O(h^p) = O(h^{p+1})$ implies that both hand-sides are zero.

Example:

$$\begin{cases} u'(x) = 4x^3 \\ u_0 = x_0^4 \end{cases}$$

We perform one integration step with the Runge–Kutta method. From (3.6) we obtain

$$\begin{cases} a_0 = 4hx_0^3 \\ b_0 = 4h(x_0 + \frac{h}{2})^3 \\ c_0 = 4h(x_0 + \frac{h}{2})^3 \\ d_0 = 4h(x_0 + h)^3. \end{cases}$$

Substituting into (3.7) and expanding in powers of h yields

$$u_1 = x_0^4 + 4x_0^3 h + 6x_0^2 h^2 + 4x_0 h^3 + h^4 = (x_0 + h)^4 = x_1^4. \qquad (3.8)$$

A proof by induction indicates that this result holds at all points.

For systems of first-order ODEs, the Runge-Kutta method is identical to (3.6)–(3.7), but now u, f, a, b, c, d are arrays. An example for a system is given in the next section.

3.6 Boundary Value Problems

Boundary value problems will be treated as particular cases of PDEs, such as one-dimensional equilibrium or steady flow problems. This type of problem can be particularly difficult to integrate using marching procedures because of singularities arising from the expression for $f(x, u)$, as can be seen with the following ODE modeling transonic flow in a converging-diverging nozzle:

$$\frac{d}{dx}\left(\frac{u^2}{2}\right) = \left(1 - \frac{u^2}{2}\right)\frac{g'}{g}, \quad 0 \le x \le 1, \tag{3.9}$$

where $g(x)$ represents the nozzle area, say $g(x) = 1 + 4(x - \frac{1}{2})^2$. u is the perturbation velocity about the sonic speed ($u = 0$). As with equation (3.3), boundary conditions may be specified both at $x = 0$ and at $x = 1$. The explicit form of this ODE is

$$u' = \frac{1 - \frac{u^2}{2}}{u}\frac{g'}{g}.$$

If the sonic condition is reached at the throat ($x = \frac{1}{2}$), the denominator goes to zero and a singularity prevents straightforward integration. Equations (3.3) and (3.9) are particular cases of mixed-type equations and will be discussed in Chaps. 8 to 11.

A classical case of boundary value problem is the self-similar incompressible viscous flow over a flat plate. The model equation is due to Blasius (1908)

$$\begin{cases} f''' + \frac{1}{2}ff'' = 0 \\ f(0) = 0 \\ f'(0) = 0 \\ f'(\infty) = 1. \end{cases} \tag{3.10}$$

This is a third-order nonlinear ODE for the unknown function $f(y)$. No exact solution is available, but very accurate numerical integrations have been performed, including Runge-Kutta integration. For the latter, three initial conditions are needed at $y = 0$. However, only two conditions are given there, corresponding to the non-slip condition, the third being the asymptotic condition that the flow become uniform at infinity (3.10.4). In order to proceed with Runge-Kutta or any marching procedure, the missing condition $f''(0) = \kappa$, the wall shear stress, is guessed and iterated upon, until the proper asymptotic behavior is satisfied. This method is called the *shooting method*. As an example of application of the shooting method and Runge-Kutta integration, equation (3.10) is transformed into a system of three first-order ODEs as

$$\begin{cases} f' = g \\ g' = h \\ h' = -\frac{1}{2}fh \end{cases}$$

with boundary conditions

$$\begin{cases} f(0) = 0 \\ g(0) = 0 \\ h(0) = \kappa. \end{cases}$$

The integration is carried out with the fourth-order Runge-Kutta method, with a step $\Delta y = 0.1$ for 100 steps. Three values of κ are used, $\kappa = 0.3,\ 0.35,\ 0.332$, the last value providing the correct asymptotic result, as seen in Fig. 3.2. The velocity profile and shear stress are shown in Fig. 3.3.

Before leaving this chapter on ODEs, it is worth stressing the unique feature of polynomials as test functions for numerical schemes. The fact that the

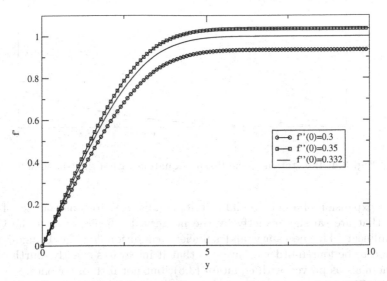

Fig. 3.2. Velocity profiles with shooting method for $\kappa = 0.3, 0.35, 0.332$

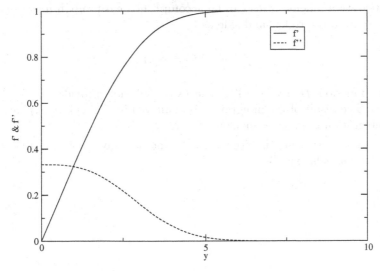

Fig. 3.3. Velocity profile and shear stress distribution for $\kappa = 0.332$

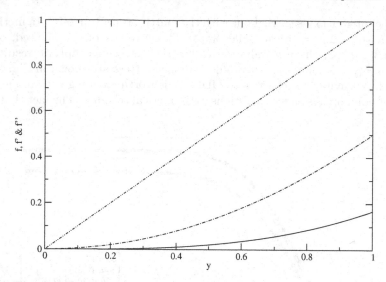

Fig. 3.4. Integration of the Blasius equation with a right-hand side

Taylor expansion of a polynomial is finite can be used to construct test functions that are satisfied exactly by the numerical scheme, even if the ODE is nonlinear. The necessary and sufficient condition for the Runge-Kutta method to be fourth-order accurate is that it integrates exactly fourth-order polynomials as proved with equation (3.8), but not fifth-order ones.

The Blasius equation (3.10) is a nonlinear ODE of third-order. Using a polynomial of high enough degree as exact solution is a non-trivial test of the integration scheme. This can be accomplished easily upon introducing a source term in the right-hand side as

$$f''' + \frac{1}{2} f f'' = g(y).$$

The function $f(y) = y^3/3!$ is the exact solution provided $g(y) = 1 + (y^4/12)$. The result of the integration is displayed in Fig. 3.4 for f, f' and f''. The integration is exact to machine accuracy.

Problem: Show that the Runge-Kutta method does not integrate fifth-order polynomials exactly.

4. Partial Differential Equations

4.1 Introduction

An ordinary differential equation (ODE) of order n has a general solution (excluding singular solutions) which depends on n arbitrary constants of integration. In the case of partial differential equations (PDE) the situation is more complicated. The general solution of a PDE does not depend on arbitrary constants, but on arbitrary functions. In general (excluding again the case of singular solutions), the number of these arbitrary functions is equal to the order of the equation. The arbitrary functions depend on one variable less than the solution itself.

Example 1: Let $u(x, y)$ be solution of

$$\frac{\partial u}{\partial y} = 0.$$

The general solution is given by $u(x, y) = f(x)$, where f is an arbitrary function of x.

Example 2:

$$\frac{\partial^2 u}{\partial x \partial y} = 0.$$

The general solution is $u(x, y) = f(x) + g(y)$.

Remark: In example 2, the second-order PDE can be transformed into an equivalent system of two first-order PDEs upon introduction of a new unknown function $v(x, y)$ as

$$\begin{cases} \dfrac{\partial u}{\partial x} = v \\[2mm] \dfrac{\partial v}{\partial y} = 0, \end{cases}$$

and the solution is obtained with two quadratures of the type of example 1.

Note that a PDE of order two is equivalent to a system of three (or two) first-order PDEs. The converse is not true. Take for example Laplace's equation

$$\frac{\partial^2 u}{\partial x^2} + \frac{\partial^2 u}{\partial y^2} = 0. \tag{4.1}$$

It is equivalent to the following system of two first-order PDEs

$$\begin{cases} \dfrac{\partial v}{\partial x} + \dfrac{\partial w}{\partial y} = 0 \\[2mm] \dfrac{\partial w}{\partial x} - \dfrac{\partial v}{\partial y} = 0, \end{cases}$$

where we have introduced $v = \partial u/\partial x$ and $w = \partial u/\partial y$.

If, instead of a zero right-hand side there is a forcing function of u (and possibly x and y), three first-order equations are required, for instance:

$$\begin{cases} \dfrac{\partial u}{\partial x} = v \\[2mm] \dfrac{\partial u}{\partial y} = w. \\[2mm] \dfrac{\partial v}{\partial x} + \dfrac{\partial w}{\partial y} = f(x, y, u) \end{cases}$$

Since it is always possible to transform a PDE of order larger than one into an equivalent system of first-order PDEs, we will limit the discussion in this chapter to systems of first-order partial differential equations.

4.2 General Classification and Notion of Characteristic Surface

Consider a first-order system of k PDEs for k unknown functions $u = \left(u^i\right)$, $i = 1, ..., k$, depending on n independent variables $x = (x_\nu)$, $\nu = 1, ..., n$. The system can be written in the form:

$$L_j(u) = a^{ij\nu} \frac{\partial u^i}{\partial x_\nu} + b^j = r^j = 0, \; j = 1, ..., k. \tag{4.2}$$

Repeated indices mean summation. The coefficients $a^{ij\nu}$ depend on x; the b^j's depend on x and also eventually on u. The following notation is introduced for short:

$$\frac{\partial u}{\partial x_\nu} = \frac{\partial u}{\partial \nu},$$

and

$$L(u) = A^\nu \frac{\partial u}{\partial \nu} + b = 0, \tag{4.3}$$

where A^ν, $\nu = 1, ..., n$ is a $k \times k$ matrix $\left(a^{ij}\right)^\nu$ and the operator L as well as b are vectors. Equation (4.3) is the *matrix form* of the system.

Consider a surface C: $\phi(x) = 0$ and $\nabla\phi \neq 0$, for example $\partial\phi/\partial n \neq 0$. On C the *characteristic matrix* is defined as:

$$A = \frac{\partial\phi}{\partial\nu} A^\nu, \qquad (4.4)$$

and the *characteristic determinant* or *characteristic form* as:

$$Q\left(\frac{\partial\phi}{\partial x_1}, ..., \frac{\partial\phi}{\partial n}\right) = |A|. \qquad (4.5)$$

We introduce the *Cauchy Problem*.

Let the initial value of the vector u (the *Cauchy data*) be given on C (the *Cauchy surface*).

The following result holds: If $Q \neq 0$ on C, then the differential system (4.3) determines in a unique way all the partial derivatives $\partial u/\partial\nu$ for arbitrary initial data. In this case, the surface C is said to be *free*. If $Q = 0$ on C, C is said to be a *characteristic surface*. In this case there exists a characteristic linear combination $\lambda L(u) = \lambda_j L_j(u) = \Lambda(u)$ of the differential operators L_j such that in Λ the derivative of the vector u on C is *interior*, i.e. depends only on u on C. $\Lambda(u) = 0$ imposes a relation between the initial data and hence these cannot be chosen arbitrarily.

The proof of the results can be found in Reference [2].

Remarks: The results are also true if the coefficients $a^{ij\nu}$ depend also on u (quasi-linear system). In this case, the characteristic condition depends on the initial data.

The characteristic equation $Q = 0$ is a PDE of first-order for $\phi(x)$. If it can be satisfied identically by a function ($\nabla\phi \neq 0$), then the family of surfaces $\phi = $ const are called *characteristic surfaces*.

As concerns the classification, if the homogeneous algebraic equation $Q = 0$ cannot be satisfied by real functions (other than $\nabla\phi = 0$) then characteristic surfaces do not exist and the system is said to be *elliptic*.

Opposite to the elliptic case, if $Q = 0$ possesses k real and distinct roots for $\partial\phi/\partial n$ for arbitrary $\partial\phi/\partial x_1, ..., \partial\phi/\partial x_{n-1}$, then the system is said to be *totally hyperbolic*.

Between the elliptic case and the totally hyperbolic case, intermediate situations are possible.

Note that a single first-order PDE is always of hyperbolic type.

A derivation of the above results that is simpler than that of Reference [2], as well as the derivation of the *compatibility relations* (or *characteristic relations*) is now presented.

Assume that $\partial\phi/\partial n \neq 0$. It is then possible to make a change of variables from $(x_1, x_2, ..., x_{n-1}, x_n)$ to $(x_1, x_2, ..., x_{n-1}, \phi)$, and $u(x_1, ..., x_\nu) = U(x_1, ..., \phi)$. The partial derivatives are modified accordingly:

$$\begin{cases} \dfrac{\partial u}{\partial \nu} = \dfrac{\partial U}{\partial \nu} + \dfrac{\partial U}{\partial \phi}\dfrac{\partial \phi}{\partial \nu}, \ \nu = 1, ..., n-1 \\[2ex] \qquad \dfrac{\partial u}{\partial n} = \dfrac{\partial U}{\partial \phi}\dfrac{\partial \phi}{\partial n} \end{cases}.$$

The system reads

$$A^\nu \frac{\partial U}{\partial \nu} + \left(\frac{\partial \phi}{\partial \nu}A^\nu\right)\frac{\partial U}{\partial \phi} + b = r = 0 \ .$$

Note that in the first term the summation runs from $\nu = 1$ to $n - 1$, whereas in the second term it runs from $\nu = 1$ to n.

The matrix combination in bracket is the characteristic matrix A and $\partial U/\partial \phi$ is an exterior derivative, i.e. points to data outside C. Rewriting the equation with a right-hand side gives

$$A\frac{\partial U}{\partial \phi} = -A^\nu \frac{\partial U}{\partial \nu} - b,$$

where the right-hand side is known because x and U are known on the Cauchy surface $\phi = 0$, and the derivatives $\partial U/\partial \nu$ are interior derivatives (derivatives holding ϕ to zero).

If $Q \neq 0$, the Cauchy problem has a unique solution.

If $Q = 0$, the Cauchy surface is a characteristic. There exists a compatibility condition on the Cauchy data. It is obtained by replacing one column of the determinant of A by the right-hand side. The right-hand side can be written $A(\partial U/\partial \phi) - r$. The determinant can now be expanded, using the linearity property and the fact that the vector $A(\partial U/\partial \phi)$ itself can be decomposed into k elements such as $A.\left(0, ..., \partial U^i/\partial \phi, ...\right)^t$, proportional to a column of A. All these determinants are zero, and there remains only the determinant with a column replaced by the vector $(-r)$. In other words, the compatibility relations are obtained by replacing a column of A by the equations themselves and setting the determinant to zero.

4.3 Model Equations and Types

4.3.1 Linear Convection Equation

The *linear convection equation* has been presented earlier (2.12) and reads

$$\frac{\partial u}{\partial t} + c\frac{\partial u}{\partial x} = r = 0.$$

Being a single first-order PDE, it is hyperbolic. Indeed, the characteristic matrix (1×1 matrix) is

$$A = \left(\frac{\partial \phi}{\partial t} + c \frac{\partial \phi}{\partial x} \right)$$

and the characteristic form

$$Q = \frac{\partial \phi}{\partial t} + c \frac{\partial \phi}{\partial x}.$$

The characteristic curve has a slope

$$\left(\frac{dx}{dt} \right)_C = -\frac{\dfrac{\partial \phi}{\partial t}}{\dfrac{\partial \phi}{\partial x}} = c.$$

The characteristics are the straight lines $C : x - ct = $ const. The compatibility relation, $r = 0$, is the equation itself. The linear convection equation represents a derivative along the characteristic (an interior derivative). If $(u)_C$ is the value of u on C, the total derivative is

$$\left(\frac{du}{dt} \right)_C = \frac{\partial u}{\partial t} + \left(\frac{dx}{dt} \right)_C \frac{\partial u}{\partial x} = \frac{\partial u}{\partial t} + c \frac{\partial u}{\partial x} = 0.$$

The total derivative is zero. Hence $(u)_C = $ const. This allows us to find the general solution to (2.12) as

$$u(x, t) = f(x - ct),$$

where f is an arbitrary function of a single variable $\xi = x - ct$.

4.3.2 The Wave Equation

In two independent variables, the *wave equation* equation reads

$$\frac{\partial^2 u}{\partial t^2} - c^2 \frac{\partial^2 u}{\partial x^2} = 0. \tag{4.6}$$

To study this equation, we transform it into a system of first-order PDEs letting $v = \partial u / \partial t$, $w = \partial u / \partial x$, to get

$$\begin{cases} \dfrac{\partial v}{\partial t} - c^2 \dfrac{\partial w}{\partial x} = r_1 = 0 \\[2mm] \dfrac{\partial w}{\partial t} - \dfrac{\partial v}{\partial x} = r_2 = 0. \end{cases} \tag{4.7}$$

Using the above notation, the matrix form of the system depends on the matrices A^t and A^x as

$$\begin{pmatrix} 1 & 0 \\ 0 & 1 \end{pmatrix} \frac{\partial}{\partial t} \begin{pmatrix} v \\ w \end{pmatrix} + \begin{pmatrix} 0 & -c^2 \\ -1 & 0 \end{pmatrix} \frac{\partial}{\partial x} \begin{pmatrix} v \\ w \end{pmatrix} = 0.$$

The characteristic matrix can be easily identified to be

$$A = \begin{pmatrix} \dfrac{\partial \phi}{\partial t} & -c^2 \dfrac{\partial \phi}{\partial x} \\[2mm] -\dfrac{\partial \phi}{\partial x} & \dfrac{\partial \phi}{\partial t} \end{pmatrix}.$$

The determinant of A is the characteristic form $Q = (\partial \phi / \partial t)^2 - c^2 (\partial \phi / \partial x)^2$. It can be factored to show the two roots

$$Q = \left(\frac{\partial \phi}{\partial t} - c \frac{\partial \phi}{\partial x} \right) \left(\frac{\partial \phi}{\partial t} + c \frac{\partial \phi}{\partial x} \right) = 0.$$

The system possesses two families of characteristic curves with slopes $(dx/dt)_{C^{\pm}} = \pm c$. The characteristics $C^{\pm}$ are the straight lines of equation $x \mp ct = \text{const}$. The wave equation is *totally hyperbolic*.

The compatibility relations are obtained from the determinant

$$\begin{vmatrix} r_1 & -c^2 \dfrac{\partial \phi}{\partial x} \\[2mm] r_2 & \dfrac{\partial \phi}{\partial t} \end{vmatrix} = 0.$$

where $\partial \phi / \partial t = \mp c (\partial \phi / \partial x)$. We obtain the two relations:

$$\begin{cases} \dfrac{\partial}{\partial t} (v - cw) + c \dfrac{\partial}{\partial x} (v - cw) = 0, \; on \; C^+ \\[3mm] \dfrac{\partial}{\partial t} (v + cw) - c \dfrac{\partial}{\partial x} (v + cw) = 0, \; on \; C^- \end{cases}.$$

These can be integrated to give:

$$\begin{cases} v - cw = f(x - ct) \\ v + cw = g(x + ct). \end{cases}$$

Integrating once more the solution for $u(x,t)$ is obtained. It is *d'Alembert's solution*

$$u(x,t) = F(x - ct) + G(x + ct), \tag{4.8}$$

F and G are arbitrary functions of a single argument.

4.3.3 Laplace's Equation

It is not necessary to do the analysis again for *Laplace's equation*

$$\frac{\partial^2 u}{\partial x^2} + \frac{\partial^2 u}{\partial y^2} = 0.$$

We use the results from the wave equation, replacing t by y and c by $\underline{i}$. The characteristic form becomes

$$Q = \left(\frac{\partial \phi}{\partial x}\right)^2 + \left(\frac{\partial \phi}{\partial y}\right)^2.$$

$Q = 0$ cannot be satisfied by a real function except the trivial solution. There does not exist real characteristic directions. Hence, Laplace's equation is elliptic.

4.3.4 The Heat Equation

The *heat equation*, equation (2.8), is now studied for its type. The solution $u(x, t)$ satisfies

$$\frac{\partial u}{\partial t} = \alpha \frac{\partial^2 u}{\partial x^2}, \ \alpha > 0.$$

Let $v = \partial u/\partial x$, then an equivalent first-order system is

$$\begin{cases} \dfrac{\partial u}{\partial t} - \alpha \dfrac{\partial v}{\partial x} = 0 \\[2mm] \dfrac{\partial u}{\partial x} = v. \end{cases}$$

In matrix form, this becomes

$$\begin{pmatrix} 1 & 0 \\ 0 & 0 \end{pmatrix} \frac{\partial}{\partial t} \begin{pmatrix} u \\ v \end{pmatrix} + \begin{pmatrix} 0 & -\alpha \\ 1 & 0 \end{pmatrix} \frac{\partial}{\partial x} \begin{pmatrix} u \\ v \end{pmatrix} = \begin{pmatrix} 0 \\ v \end{pmatrix}.$$

The characteristic matrix follows as

$$A = \begin{pmatrix} \dfrac{\partial \phi}{\partial t} & -\alpha \dfrac{\partial \phi}{\partial x} \\[3mm] \dfrac{\partial \phi}{\partial x} & 0 \end{pmatrix},$$

and the characteristic form as $Q = -\alpha \left(\partial \phi/\partial x\right)^2$. The characteristic form is degenerated because it does not contain the derivative $\partial \phi/\partial t$. $Q = 0$ is satisfied by $\partial \phi/\partial x = 0$ and $\partial \phi/\partial t = 1$ (arbitrary), which is a double root. The characteristic lines are the straight lines $t = $ const. Only one compatibility relation is found.

The heat equation is parabolic.

Note that the slope of the characteristic is $(dx/dt)_C = \pm\infty$, which is the property of infinite speed of propagation in both directions on the x-axis (double root).

4.3.5 Burgers' Equation (Inviscid)

Burgers' equation reads

$$\frac{\partial u}{\partial t} + \frac{\partial}{\partial x}\left(\frac{u^2}{2}\right) = r = 0. \tag{4.9}$$

For the analysis of the type, the solution is assumed to be smooth locally, so that the quasi-linear form can be used:

$$\frac{\partial u}{\partial t} + u\frac{\partial u}{\partial x} = 0.$$

The characteristic matrix is $A = ((\partial\phi/\partial t) + u(\partial\phi/\partial x))$, and the characteristic form

$$Q = |A| = \frac{\partial\phi}{\partial t} + u\frac{\partial\phi}{\partial x} = 0, \tag{4.10}$$

has a root corresponding to the characteristic direction $(dx/dt)_C = u$.

The compatibility relation is $r = 0$, the equation itself. It states that if $(u)_C$ is the value of u along the characteristic, $(du_C/dt) = (\partial u/\partial t) + (dx/dt)_C(\partial u/\partial x) = 0$, hence u is constant on C. The slope of C is also constant as a consequence, hence the characteristics are the straight lines $\xi = x - ut = \text{const}$.

Burgers equation is hyperbolic. The general solution to Burgers' equation can be written in the following implicit form

$$u(x,t) = F(x - ut),$$

where $F(\xi)$ is an arbitrary function of a single argument. Depending on the initial and boundary conditions which determine the functional form of F, an explicit solution for u may be obtained. An example of this is the expansion solution which corresponds to the initial value problem

$$\begin{cases} \dfrac{\partial u}{\partial t} + \dfrac{\partial}{\partial x}\left(\dfrac{u^2}{2}\right) = 0 \\ u(x,0) = x, \quad -\infty < x < \infty. \end{cases}$$

We apply the initial condition to the general form of the solution and get

$$u(x,0) = F(x) = x, \quad -\infty < x < \infty.$$

This yields $F(\xi) = \xi$, $-\infty < \xi < \infty$. Now, replacing ξ by $x - ut$ allow us to solve for u as

$$u(x,t) = \frac{x}{1+t}.$$

4.3.6 Other Examples

i) $\Delta^2 u = 0$, i.e. $\sum_{i,j=1}^{n}(\partial^4 u/\partial x_i^2 \partial x_j^2) = 0$. The characteristic form is

$$Q = \left(\sum_{i=1}^{n}\left(\frac{\partial\phi}{\partial x_i}\right)^2\right)^2.$$

The equation is of elliptic type.

ii) $\sum_{i=1}^{n}(\partial^4 u/\partial x_i^4) = 0$, has the following characteristic form:

$$Q = \sum_{i=1}^{n}\left(\frac{\partial\phi}{\partial x_i}\right)^4.$$

The equation is elliptic.

iii) The operator $\left(\Delta - (\partial^2/\partial t^2)\right)\left(\Delta - 2(\partial^2/\partial t^2)\right)u = 0$ is hyperbolic since its characteristic form is

$$Q = \left(\sum_{i=1}^{n}\left(\frac{\partial\phi}{\partial x_i}\right)^2 - \left(\frac{\partial\phi}{\partial t}\right)^2\right)\left(\sum_{i=1}^{n}\left(\frac{\partial\phi}{\partial x_i}\right)^2 - 2\left(\frac{\partial\phi}{\partial t}\right)^2\right).$$

iv) The operator $\left(\Delta - (\partial^2/\partial t^2)\right)\left(\Delta + (\partial^2/\partial t^2)\right)u = 0$ is of intermediary type. It is not elliptic, nor parabolic, nor hyperbolic. Its characteristic form is:

$$Q = \left(\sum_{i=1}^{n}\left(\frac{\partial\phi}{\partial x_i}\right)^2\right)^2 - \left(\frac{\partial\phi}{\partial t}\right)^4.$$

Problem: Prove the statements in this article by establishing the characteristic forms of each of the given model PDEs.

4.4 Conservation Laws and Jumps
for a System of PDEs

Let $u = (u^i)$, $i = 1, ..., k$ be the dependent variables and $x = (x_\nu)$, $\nu = 1, ..., n$ be the independent variables. A system of partial differential equations is said to be in *divergence* or *conservation law form* if each equation can be formulated as the divergence of a vector $v(u)$ as

$$L_j(u) = \frac{\partial v^{j\nu}(u)}{\partial \nu} + b^j = 0,$$

or for short

$$L(u) = \frac{\partial v^\nu}{\partial \nu} + b = 0. \tag{4.11}$$

Examples:

Burgers equation $(\partial u/\partial t) + (\partial/\partial x)\left(\frac{u^2}{2}\right) = 0$ is in conservation form. The quasi-linear form $(\partial u/\partial t) + u(\partial u/\partial x) = 0$ is not a conservation form.

The 1-D Euler equations in conservation form read:

$$\begin{cases} \dfrac{\partial \rho}{\partial t} + \dfrac{\partial \rho u}{\partial x} = 0 \\[2mm] \dfrac{\partial \rho u}{\partial t} + \dfrac{\partial(\rho u^2 + p)}{\partial x} = 0 \\[2mm] \dfrac{\partial \rho E}{\partial t} + \dfrac{\partial \rho u H}{\partial x} = 0, \end{cases} \qquad (4.12)$$

where the total energy is $E = (p/(\gamma - 1)\rho) + (u^2/2)$, and the total enthalpy $H = (\gamma p/(\gamma - 1)\rho) + (u^2/2)$. If the last equation (4.12) is replaced by $(\partial \rho s/\partial t) + (\partial \rho s u/\partial x) = 0$, where s is the entropy, $s = C_v \ln(p/\rho^\gamma)$, the system is also in conservation form, but will not give the same solution if discontinuities are present, because the two systems have different jump conditions. However, if the solution is smooth everywhere, the solutions will be identical.

4.4.1 Jump Conditions

The system of PDEs (4.11) can be integrated on a control volume, CV bounded by the control surface CS:

$$\int_{CV} \left(\frac{\partial v^\nu(u)}{\partial \nu} + b\right) dV = 0,$$

and transformed using Gauss' theorem (divergence theorem) as

$$\int_{CS} v^\nu(u).n_\nu dA + \int_{CV} b dV = 0. \qquad (4.13)$$

The derivatives have disappeared. The conditions for the existence of the last two integrals are less restrictive than for the differential form, since all that is required is that the flux vector $v(u)$ and the source term b be integrable. In particular, if $v(u)$ has jumps in the control volume, the derivatives are not defined there, but the surface integral is well defined.

The jump conditions associated with the conservation laws (4.11) are found now; Assume that $v(u)$ is discontinuous along a surface S in the control volume, such that S divides CV in two parts, CV_1 and CV_2, and the control surface in CS_1 and CS_2, then applying (4.13) to parts 1 and 2 yields:

$$\int_{CS_1+S} v^\nu(u).n_\nu dS + \int_{CV_1} b dV = 0$$

$$\int_{CS_2+S} v^\nu(u).n_\nu dS + \int_{CV_2} bdV = 0.$$

The integral form of the conservation law (4.13) is also valid when discontinuities are present inside the control volume, hence

$$\int_{CS_1+CS_2} v^\nu(u).n_\nu dS + \int_{CV_1+CV_2} bdV = 0.$$

Subtracting the previous two equations from the last one yields:

$$\int_S (v_1^\nu.n_\nu^1 + v_2^\nu.n_\nu^2)dS = 0.$$

Let $\langle a \rangle = a_2 - a_1$ denote the jump of a, and $n_\nu = n_\nu^2 = -n_\nu^1$. The last result takes the form

$$\int_S \langle v^\nu \rangle .n_\nu dS = 0.$$

This equation must hold for any part of the surface S, hence the integrand is zero, i.e.

$$\langle v^\nu \rangle .n_\nu = 0. \tag{4.14}$$

These are the jump conditions for a system of conservation laws.

4.4.2 Examples

i) Jump condition for Burgers' equation $(\partial u/\partial t)+(\partial/\partial x)(u^2/2)=0$.
By identification we find $v^t = u$ and $v^x = \frac{u^2}{2}$. (4.14) gives

$$\langle u \rangle n_t + \left\langle \frac{u^2}{2} \right\rangle n_x = 0.$$

The jump-line or shock has a slope

$$\left(\frac{dx}{dt} \right)_S = -\frac{n_t}{n_x} = \frac{\left\langle \dfrac{u^2}{2} \right\rangle}{\langle u \rangle} = \frac{u_1 + u_2}{2},$$

where the indices refer to the states on each side of the shock.
The following initial data

$$\begin{cases} u(x,0) = 1, \ x < 0 \\ u(x,0) = 0, \ x > 0. \end{cases}$$

produce a shock S moving at constant speed $(dx/dt)_S = \frac{1}{2}$. In Fig. 4.1 the arrows represent a characteristic vector field $\vec{V} = (u(x,t),1)$. The characteristic relation, equation (4.10) can be interpreted as a vector dot product,

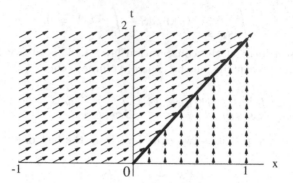

Fig. 4.1. Moving shock with Burgers' equation

$\vec{V}.\nabla\phi = 0$, stating that the characteristic lines are tangent to the vector field $\vec{V}$.

If the initial conditions are reversed, i.e.

$$\begin{cases} u(x,0) = 0, \ x < 0 \\ u(x,0) = 1, \ x > 0. \end{cases}$$

an expansion shock is ruled out, as violating the condition that characteristics must originate from the initial condition, and cannot appear in the middle of the domain. The correct solution is an expansion fan, as displayed in Fig. 4.2. The analytic form of the solution reads:

$$\begin{cases} u = \dfrac{x}{t}, \ 0 \le \dfrac{x}{t} \le 1 \\ u = 0, \ \dfrac{x}{t} \le 0 \\ u = 1, \ \dfrac{x}{t} \ge 1. \end{cases}$$

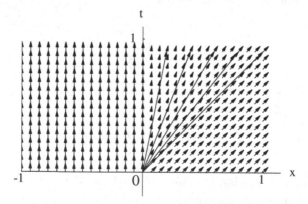

Fig. 4.2. Expansion fan with Burgers' equation

ii) 1-D Euler Equations. The jump conditions associated with (4.12) are given by:

$$\left\langle \begin{array}{c} \rho \\ \rho u \\ \rho E \end{array} \right\rangle n_t + \left\langle \begin{array}{c} \rho u \\ \rho u^2 + p \\ \rho u H \end{array} \right\rangle n_x = 0.$$

For steady flow ($n_t = 0$) the conditions reduce to

$$\left\langle \begin{array}{c} \rho u \\ \rho u^2 + p \\ \rho u H \end{array} \right\rangle = 0,$$

which can be further simplified by taking into account the first equation to reduce the last jump condition to $\langle H \rangle = 0$. The total enthalpy is conserved across a steady shock.

iii) Elliptic System. Consider the system of first-order PDEs (4.7) which, for $t = y$ and $c = \underline{i}$ corresponds to Laplace's equation

$$\begin{cases} \dfrac{\partial w}{\partial x} + \dfrac{\partial v}{\partial y} = 0 \\[2mm] \dfrac{\partial v}{\partial x} - \dfrac{\partial w}{\partial y} = 0. \end{cases} \tag{4.15}$$

The jump conditions are

$$\left\langle \begin{array}{c} w \\ v \end{array} \right\rangle n_x + \left\langle \begin{array}{c} v \\ -w \end{array} \right\rangle n_y = 0.$$

The elimination of n_x and n_y yields

$$\langle v \rangle^2 + \langle w \rangle^2 = 0,$$

which cannot be satisfied, except by the trivial solution $\langle v \rangle = \langle w \rangle = 0$. Laplace's equation does not admit jumps.

Problem: Prove that the linear convection equation admits jumps that are arbitrary, along characteristic lines.

5. Integration of a Linear Hyperbolic Equation

5.1 Introduction

Hyperbolic PDEs are associated with wave phenomena and propagation. A vibrating string, a vibrating column of air in a wind instrument, supersonic flow and shock waves, are some examples of physical phenomena described by hyperbolic equations. We will limit ourselves in this chapter to one-dimensional unsteady and two-dimensional steady model problems, governed by the linear convection equation and the wave equation respectively. The two model equations have been studied in the previous chapter and shown to be of hyperbolic type.

5.2 The Linear Convection Equation

Consider the linear convection equation for the unknown $u(x, t)$

$$\frac{\partial u}{\partial t} + c\frac{\partial u}{\partial x} = 0, \ c > 0,$$

where the convection speed c can be made positive by an appropriate choice of direction for the x-axis. In order to advance the solution in time, the new value, u_i^{n+1} of the solution must appear in the time derivative. The scheme for $\partial u/\partial t$ will therefore be an advanced scheme of the type (2.1). Several classical schemes are presented and analyzed in the next paragraphs.

5.2.1 A Centered Scheme

The space derivative is approximated by the centered FD scheme (2.3)

$$\frac{u_i^{n+1} - u_i^n}{\Delta t} + c\frac{u_{i+1}^n - u_{i-1}^n}{2\Delta x} = 0. \tag{5.1}$$

This scheme is first-order accurate in time and second-order accurate in space as indicated by the TE

$$\epsilon_i^n = \frac{\Delta t}{2}\frac{\partial^2 u_i^n}{\partial t^2} + c\frac{\Delta x^2}{6}\frac{\partial^3 u_i^n}{\partial x^3} + O(\Delta t^2, \Delta x^4) = O(\Delta t, \Delta x^2).$$

As far as stability is concerned, this scheme is unconditionally unstable, therefore it is useless. Substituting $u_i^n = g^n e^{iia}$ into the equation with $\sigma = c(\Delta t / \Delta x)$ and solving for the amplification factor yields

$$g = 1 - i\sigma \sin \alpha.$$

$|g|$ is always larger than one for all non zero σ.

Although this is not a good scheme for this equation, it is always recommended, when a first attempt is made at solving a PDE, to try the simplest schemes first. But one must not expect that it will necessarily work. We will see that this scheme will play an important role in Burgers' equation (4.9) at a shock. The nonlinearity is the reason why it can be used there.

5.2.2 An Upwind Scheme

We use scheme (2.2) which is qualified as retarded or *upwind*:

$$\frac{u_i^{n+1} - u_i^n}{\Delta t} + c\frac{u_i^n - u_{i-1}^n}{\Delta x} = 0. \tag{5.2}$$

This scheme is consistent and first-order accurate in x and t. The TE reads

$$\epsilon_i^n = \frac{\Delta t}{2}\frac{\partial^2 u_i^n}{\partial t^2} - c\frac{\Delta x}{2}\frac{\partial^2 u_i^n}{\partial x^2} + O(\Delta t^2, \Delta x^2) = O(\Delta t, \Delta x).$$

One of the second partial derivatives can be eliminated using the identity obtained from the equation as

$$\frac{\partial^2 u}{\partial t^2} = \frac{\partial}{\partial t}\left(-c\frac{\partial u}{\partial x}\right) = -c\frac{\partial}{\partial x}\left(\frac{\partial u}{\partial t}\right) = c^2\frac{\partial^2 u}{\partial x^2}.$$

The TE is modified accordingly. If we use the same notation for σ, the result is

$$\epsilon_i^n = c\frac{\Delta x}{2}(\sigma - 1)\frac{\partial^2 u_i^n}{\partial x^2} + O(\Delta t^2, \Delta x^2) = O\left((1 - \sigma)\Delta x\right).$$

When $\sigma = 1$ the scheme is of higher accuracy than one. In fact it is exact, i.e. one can show that $\epsilon_i^n = 0$. This seems intriguing, but there is a simple explanation. First note that equation (5.2) reduces in this case to $u_i^{n+1} = u_{i-1}^n$. Also, notice that $\xi_i^{n+1} = x_i - ct^{n+1} = x_{i-1} - ct^n = \xi_{i-1}^n$, indicating that the two points $(i-1, n)$ and $(i, n+1)$ are on the same characteristic line. Since the exact solution is $u(x, t) = F(x - ct) = F(\xi)$ the numerical scheme is transporting the constant value of u along the characteristic, as shown in Fig. 5.1.

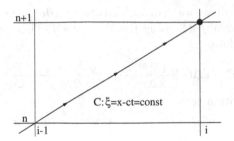

Fig. 5.1. Computational molecule and characteristic line for $\sigma = 1$

Let $u_i^n = g^n e^{ii\alpha}$. One finds

$$g = 1 - \sigma(1 - \cos\alpha) - \underline{i}\sigma\sin\alpha,$$

and

$$|g|^2 = 1 - 2\sigma(1 - \sigma)(1 - \cos\alpha).$$

Stability requires $|g| \leq 1$, which is verified when the condition $0 \leq \sigma \leq 1$ is met, the CFL condition (2.14).

Note that for negative values of c, an advanced scheme would be needed.

5.2.3 The Lax Scheme

The Lax scheme has been introduced earlier:

$$\frac{u_i^{n+1} - \frac{u_{i-1}^n + u_{i+1}^n}{2}}{\Delta t} + c\frac{u_{i+1}^n - u_{i-1}^n}{2\Delta x} = 0.$$

We only report the results here. Lax scheme is conditionally consistent (2.13)

$$\epsilon_i^n = \frac{\Delta t}{2}\frac{\partial^2 u_i^n}{\partial t^2} - \frac{\Delta x^2}{2\Delta t}\frac{\partial^2 u_i^n}{\partial x^2} + O\left(\Delta x^2, \frac{\Delta x^4}{\Delta t}\right) = O\left(\Delta t, \frac{\Delta x^2}{\Delta t}, \Delta x^2\right),$$

and is stable under the CFL condition (2.14)

$$\Delta t \leq \frac{\Delta x}{|c|}.$$

5.2.4 The Lax-Wendroff Scheme

This scheme is obtained using a systematic procedure. Consider the Taylor expansion of u_i^{n+1}:

$$u_i^{n+1} = u_i^n + \Delta t\frac{\partial u_i^n}{\partial t} + \frac{\Delta t^2}{2}\frac{\partial^2 u_i^n}{\partial t^2} + O(\Delta t^3).$$

We have seen how we can substitute $-c(\partial u_i^n/\partial x)$ for $\partial u_i^n/\partial t$ and $c^2(\partial^2 u_i^n/\partial x^2)$ for $\partial^2 u_i^n/\partial t^2$, using the PDE itself. This yields the following scheme in update form

$$u_i^{n+1} = u_i^n - c\Delta t\frac{u_{i+1}^n - u_{i-1}^n}{2\Delta x} + c^2\frac{\Delta t^2}{2}\frac{u_{i+1}^n - 2u_i^n + u_{i-1}^n}{\Delta x^2},$$

which can be rewritten as

$$\frac{u_i^{n+1} - u_i^n}{\Delta t} + c\frac{u_{i+1}^n - u_{i-1}^n}{2\Delta x} - c^2\frac{\Delta t}{2}\frac{u_{i+1}^n - 2u_i^n + u_{i-1}^n}{\Delta x^2} = 0. \qquad (5.3)$$

A Taylor expansion about point (i, n) gives

$$\epsilon_i^n = \frac{\Delta t^2}{3!}\frac{\partial^3 u_i^n}{\partial t^3} + c\frac{\Delta x^2}{3!}\frac{\partial^3 u_i^n}{\partial x^3} + O(\Delta t\Delta x^2, \Delta t^3, \Delta x^4).$$

Again, it is possible to eliminate one of the third derivatives, since $\partial^3 u/\partial t^3 = -c^3(\partial^3 u/\partial x^3)$, to get

$$\epsilon_i^n = c\frac{\Delta x^2}{3!}(1 - \sigma^2)\frac{\partial^3 u_i^n}{\partial x^3} + O(\Delta t\Delta x^2, \Delta t^3, \Delta x^4) = O\left((1 - \sigma^2)\Delta x^2\right).$$

Lax–Wendroff scheme is second-order accurate in t and x.

Note that the scheme reduces to $u_i^{n+1} = u_{i-1}^n$ when $\sigma = 1$. The exact solution is obtained ($\epsilon_i^n = 0$).

Problem: When $\sigma^2 \neq 1$, look for the point about which the Taylor expansion does not contain the term $\Delta t\Delta x^2$.

The amplification factor for the complex mode is

$$g = 1 - \sigma^2(1 - \cos\alpha) - i\sigma\sin\alpha.$$

Computing the square of the modulus of the amplification factor as

$$|g|^2 = \left(1 - \sigma^2(1 - \cos\alpha)\right)^2 + \sigma^2\sin^2\alpha = 1 - \sigma^2(1 - \sigma^2)(1 - \cos\alpha),$$

which is less than one when $|\sigma| \leq 1$. The absolute value has been introduced because Lax-Wendroff scheme is symmetric and can accept positive or negative convection speeds.

5.2.5 The MacCormack Scheme

In 1969 MacCormack introduced a new two-step predictor-corrector scheme. The overbarred values correspond to the predictor values:

$$\overline{u_i^{n+1}} = u_i^n - c\Delta t\frac{u_{i+1}^n - u_i^n}{\Delta x}$$

$$u_i^{n+1} = \frac{1}{2}\left(u_i^n + \overline{u_i^{n+1}} - c\Delta t\frac{\overline{u_i^{n+1}} - \overline{u_{i-1}^{n+1}}}{\Delta x}\right).$$

For a linear equation, such as the linear convection equation, this scheme is identical to the Lax-Wendroff scheme. Indeed, if one eliminates the predictor values in the corrector step, one obtains (5.3).

5.3 The Wave Equation

Consider the wave equation (4.6). It is a second-order PDE; it models the vibrations of a string for a string instrument or of a column of air for a wind instrument. If we exchange the variable t for y and c for β, it can be interpreted as the equation governing a supersonic stream of air, moving parallel and in the direction of the x-axis; the incoming flow is undisturbed upstream of the airfoil and is tangent to the surface of the airfoil, a slender wedge of half-angle θ, placed along the x-axis. The equation reads

$$-\beta^2 \frac{\partial^2 \varphi}{\partial x^2} + \frac{\partial^2 \varphi}{\partial y^2} = 0, \ \beta = \sqrt{M_0^2 - 1}, \ M_0 \geq 1. \tag{5.4}$$

This is the equation for the velocity potential for linearized supersonic flow. φ is the velocity potential for the perturbation. M_0 is the incoming flow *Mach number*. The flow field is depicted in Fig. 5.2.

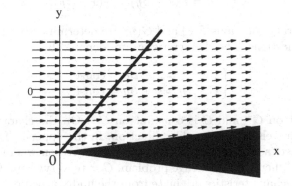

Fig. 5.2. Supersonic flow past a slender wedge

Note that unlike (4.6) both independent variables represent space. Therefore, depending on the problem, in particular the boundary conditions, one of the variables will be considered time-like and the solution will be integrated in the direction of that variable. We are interested in finding the solution of the supersonic flow past a given geometry and no boundary condition is required downstream of the profile; x is time-like and the solution is marched in the flow direction. An example of y being time-like is the design of a supersonic diffuser producing a uniform supersonic exit flow.

For the wedge, the initial condition is

$$\varphi(x, y) = 0, \ x \leq 0, \tag{5.5}$$

and the boundary condition for the upper-half plane

$$\frac{\partial \varphi}{\partial y}(x, 0) = \theta, \ x \geq 0. \tag{5.6}$$

The velocity components are

$$\begin{cases} u = u_0 \left(1 + \dfrac{\partial \varphi}{\partial x}\right) \\ v = u_0 \dfrac{\partial \varphi}{\partial y} \end{cases}.$$

u_0 is the incoming flow velocity. In the mathematical model, the tangency condition on the wedge (5.6) is applied on the x-axis, which is justified, as a first-order approximation, if the wedge angle is small.

5.3.1 Exact Solution

We use the general solution to the wave equation to solve the flow past the wedge. Recall equation (4.8) with the current notation

$$\varphi(x, y) = F(x - \beta y) + G(x + \beta y).$$

The arbitrary functions $F(\xi)$ and $G(\eta)$ are determined by the initial and boundary conditions. The characteristics are denoted by

$$\begin{cases} C^+ : \xi = x - \beta y = \text{const} \\ C^- : \eta = x + \beta y = \text{const} \end{cases}.$$

F is constant on C^+ and G is constant on C^-. On the characteristics the information travels in the time-like direction. In the upper-half plane all the C^- characteristics originate from the undisturbed region where $F = G = 0$, equation (5.5), hence in the wedge problem, $G = 0$ everywhere. On the other hand, the C^+ characteristics originate from the undisturbed region for $\xi \leq 0$, but they originate at the wedge for $\xi \geq 0$, so that $F = 0$ for $\xi \leq 0$, and F will be obtained from equation (5.6) for $\xi \geq 0$ (see Fig. 5.3).

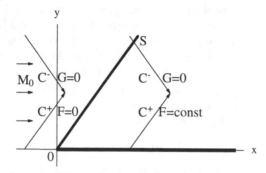

Fig. 5.3. D' Alembert method of solution

The tangency condition gives

$$\frac{\partial \varphi}{\partial y}(x,0) = -\beta F'(x) = \theta, \ x \geq 0.$$

This is an ODE for F. It can be integrated as $F(x) = -(\theta/\beta)x + \text{const}$. The integration constant is zero by continuity since $\varphi(0,0) = 0$. Now that we know the functional form of F, that is $F(\xi) = -(\theta/\beta)\xi$, $\xi \geq 0$, the solution is known in the entire upper plane $\xi \geq 0$, $y \geq 0$

$$\varphi(x,y) = -\frac{\theta}{\beta}(x - \beta y).$$

It can be easily verified that this function verifies the PDE and the other conditions. The velocity field is uniform with components

$$\begin{cases} u = u_0 \left(1 - \frac{\theta}{\beta}\right) \\ v = u_0 \theta \end{cases}.$$

The characteristic $S : \xi = 0$ plays the role of a shock. The flow properties are discontinuous there. $\langle u \rangle = -u_0(\theta/\beta)$, $\langle v \rangle = u_0\theta$ satisfy the jump conditions associated with the first-order system.

Note that since there is no reference length in this problem, the solution can be written in self-similar form as

$$f(t) = -\beta \frac{\varphi(x,y)}{\theta x} = 1 - \beta \frac{y}{x} = 1 - t.$$

Problem: Verify that the jumps, $\langle u \rangle = -u_0(\theta/\beta)$, $\langle v \rangle = u_0\theta$, satisfy the jump conditions associated with the PDE.

5.3.2 Numerical Scheme, Consistency and Accuracy

The numerical solution of equations (5.4)–(5.6) by finite differences requires the construction of a mesh system for the discretization of the PDE and of the initial and boundary conditions. The discrete solution will be defined at the nodes of the mesh. We use a Cartesian mesh with constant steps $x_i = (i-1)\Delta x$, $i = 1, 2, ...$, $y_j = (j-1)\Delta y$, $j = 1, 2, ...$ as shown in Fig. 5.4.

We introduce the following scheme for (5.4)

$$-\beta^2 \frac{\varphi_{i+1,j} - 2\varphi_{i,j} + \varphi_{i-1,j}}{\Delta x^2} + \frac{\varphi_{i,j+1} - 2\varphi_{i,j} + \varphi_{i,j-1}}{\Delta y^2} = 0, \ i = 1, ..., j = 2, ...$$

$$(5.7)$$

The computational molecule is displayed in Fig. 5.4.

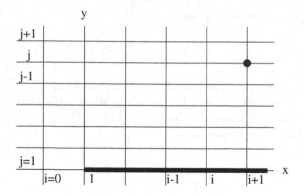

Fig. 5.4. Mesh system and computational molecule

The analysis of the TE of this centered scheme is straightforward

$$\epsilon_{i,j} = -\beta^2 \left(\frac{\partial^2 \varphi_{i,j}}{\partial x^2} + \frac{\Delta x^2}{4!} \frac{\partial^4 \varphi_{i,j}}{\partial x^4} + O(\Delta x^4) \right) + \frac{\partial^2 \varphi_{i,j}}{\partial y^2} + \frac{\Delta y^2}{4!} \frac{\partial^4 \varphi_{i,j}}{\partial y^4} + O(\Delta y^4)$$

$$= O(\beta^2 \Delta x^2, \Delta y^2).$$

The scheme is consistent and second-order accurate in x and y.

The initial condition (5.5) is implemented by extending the domain to $x_0 = -\Delta x$ and setting the values of $\varphi_{0,j}$ and $\varphi_{1,j}$ to zero

$$\begin{cases} \varphi_{0,j} = 0 \\ \varphi_{1,j} = 0 \end{cases}, j = 1, \ldots \tag{5.8}$$

This enforces not only $\varphi = 0$, but also $\partial \varphi / \partial x = 0$. No error is introduced through this initial condition.

The FDE for the boundary condition is

$$\frac{\varphi_{i+1,2} - \varphi_{i+1,1}}{\Delta y} = \theta, \, i = 1, \ldots \tag{5.9}$$

The computational molecule for the application of the boundary condition is shown in Fig. 5.5.

Here it is convenient to expand the TE about point $(i + 1, 1)$ as

$$\epsilon_{i+1,1} = \frac{\partial \varphi_{i+1,1}}{\partial y} + \frac{\Delta y}{2} \frac{\partial^2 \varphi_{i+1,1}}{\partial y^2} + O(\Delta y^2) - \theta = O(\Delta y). \tag{5.10}$$

This is a first-order accurate scheme.

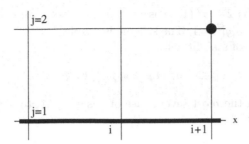

Fig. 5.5. Computational molecule for the boundary condition

5.3.3 Numerical Implementation

As mentioned, there is no reference length in this problem. Without restriction one can choose Δy such that $\theta \Delta y = 1$. Let $\sigma = \Delta x / \beta \Delta y$. The marching procedure consists in implementing the initial condition (5.8) along the two columns at $x_0 = -\Delta x$ and $x_1 = 0$. Then, scheme (5.7) is used to solve for the new values $\varphi_{i+1,j}$, $j = 2, ...$, followed by the application of the boundary condition (5.9) to obtain $\varphi_{i+1,1}$. In update form this reads

$$\begin{cases} \varphi_{i+1,j} = 2(1-\sigma^2)\varphi_{i,j} - \varphi_{i-1,j} + \sigma^2(\varphi_{i,j+1} + \varphi_{i,j-1}), \ j = 2, ... \\ \varphi_{i+1,1} = \varphi_{i+1,2} - 1. \end{cases} \quad (5.11)$$

Problem: Solve the above equations and plot the self-similar solution $-\beta(\varphi_{i,j}/\theta x_i)$ as a function of $t = \beta(y_j/x_i)$, in the three cases, $\sigma = 1/\sqrt{2}$, $\sqrt{2}$, 1. For $\sigma = 1$ prove that the scheme computes the exact solution.

5.3.4 Stability

Let $\varphi_{i,j} = g^i e^{ij\alpha}$. This choice is based on the fact that x is the marching direction (time-like) and y is space-like. Substituting this in the first of equations (5.11) gives after some simplification

$$g = 2(1-\sigma^2) - \frac{1}{g} + \sigma^2(e^{i\alpha} + e^{-i\alpha}).$$

Rearranging terms yields a quadratic equation with real coefficients for g:

$$f(g) = g^2 - 2\left(1 - \sigma^2\left(1 - \cos\alpha\right)\right)g + 1 = 0.$$

Two cases are possible:

i) $D^2 = \left(1 - \sigma^2\left(1 - \cos\alpha\right)\right)^2 - 1 = -\sigma^2\left(1 - \cos\alpha\right)\left(2 - \sigma^2\left(1 - \cos\alpha\right)\right) \geq 0$, or equivalently $2 - \sigma^2\left(1 - \cos\alpha\right) \leq 0$. In this case, the roots g_1, g_2 are real, but since $f(1) \geq 0$ and $f(-1) \leq 0$, one of the roots is outside the interval $[-1, 1]$ and $|g| \geq 1$; the scheme is unstable,

ii) $D^2 \leq 0$, that is $2 - \sigma^2 (1 - \cos \alpha) \geq 0$, the roots are complex conjugate and verify $|g_1 g_2| = |g|^2 = 1$. The scheme is stable.

The condition of stability is

$$2 - \sigma^2 (1 - \cos \alpha) \geq 0, \ \forall \alpha,$$

which reduces, in the most severe case of $\cos \alpha = -1$, to $\sigma \leq 1$. In terms of the marching step this is

$$\Delta x \leq \beta \Delta y,$$

a CFL condition for the "time step" Δx, for a given Δy.

Remark: When $\sigma = 1$, the TE is identically zero. The scheme reads

$$\varphi_{i+1,j} = -\varphi_{i-1,j} + \varphi_{i,j+1} + \varphi_{i,j-1}.$$

Note first that the characteristics coincide with the diagonals of the mesh system. If we assume that the values $\varphi_{i,j}$'s in the right-hand side are exact, then, using d'Alembert solution

$$\phi_{i+1,j} = -F_{i-1,j} - G_{i-1,j} + F_{i,j+1} + G_{i,j+1} + F_{i,j-1} + G_{i,j-1}.$$

However, from the exact solution

$$\begin{cases} F_{i,j+1} = F_{i-1,j} \\ G_{i,j-1} = G_{i-1,j} \\ F_{i+1,j} = F_{i,j-1} \\ G_{i+1,j} = G_{i,j+1}, \end{cases}$$

hence

$$\varphi_{i+1,j} = F_{i+1,j} + G_{i+1,j},$$

which is the exact solution. Since the values of $\varphi_{0,j}$ and $\varphi_{1,j}$ are exact, it follows that all the values in the triangle bounded by $\xi = 0$ and $\eta = \eta_{max}$ will be exact (zero in the present problem). Furthermore, for the wedge problem the solution in the disturbed region varies linearly with x and y and the TE for the boundary condition (Equation (5.10)) vanishes as well in this case. The exact solution is obtained everywhere in the upper-plane.

The geometric interpretation of the above condition has to do with the relative positions of the characteristics of the PDE and the diagonal lines of the mesh system. The latter have a slope

$$\left(\frac{dy}{dx} \right)_{N^\pm} = \pm \frac{\Delta y}{\Delta x},$$

whereas the former have a slope

$$\left(\frac{dy}{dx} \right)_{C^\pm} = \pm \frac{1}{\beta}.$$

The diagonal lines will be called, by extension, the *numerical character-istics*. They reflect the manner by which the information propagates in the mesh.

The stability condition can be stated as

$$\left| \left(\frac{dy}{dx} \right)_{C\pm} \right| \leq \left| \left(\frac{dy}{dx} \right)_{N\pm} \right|. \tag{5.12}$$

The stability condition requires that the numerical domain of dependence contains the domain of dependence of the PDE. This is illustrated in Fig. 5.6 showing a stable situation.

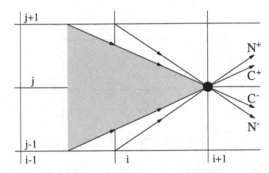

Fig. 5.6. Domains of dependence of the numerical scheme and of the PDE

Remark: Scheme (5.7) is called an *explicit* scheme, because the new values $\varphi_{i+1,j}$ can be computed explicitly from the FDEs (5.11). In general, an explicit scheme for the wave equation is subject to a stability condition of CFL type. If β is of order unity, the stability restriction is not too severe. However, if β is very small (i.e. $M_0 \cong 1$), Δx may be prohibitively small, smaller than strictly required for accuracy, since

$$\epsilon_{i,j} = O(\beta^2 \Delta x^2, \Delta y^2) = O(\beta^4 \Delta y^2, \Delta y^2) = O(\Delta y^2).$$

Δy is governing the accuracy, which is not improved by the fact that Δx is very small. The computing time can become important as many steps are necessary to reach a given value of x. The alternative to such a situation is to use an *implicit* scheme.

5.4 Implicit Scheme for the Wave Equation

Consider the following scheme

$$\begin{cases} -\beta^2 \dfrac{\varphi_{i+1,j} - 2\varphi_{i,j} + \varphi_{i-1,j}}{\Delta x^2} + \dfrac{\varphi_{i+1,j+1} - 2\varphi_{i+1,j} + \varphi_{i+1,j-1}}{\Delta y^2} = 0, \\ \qquad\qquad i = 1, ..., \ j = 2, ... \end{cases} \quad (5.13)$$

with the same initial and boundary conditions as before.

The computational molecule is shown in Fig. 5.7. It can be seen that for this scheme the numerical characteristics have slopes $(dy/dx)_{N\pm} = \pm\infty$. Condition (5.12) is satisfied for all values of β.

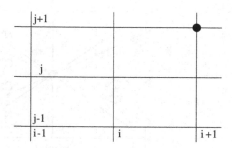

Fig. 5.7. Computational molecule for the implicit scheme

Equation (5.13) cannot be solved pointwise since it contains three unknowns at the new level $i+1$. This scheme requires the simultaneous solution of the values of φ along a column. Written as a linear algebraic system, the associated matrix is tridiagonal and can be solved easily using the Thomas algorithm. Such a situation will be discussed in detail in Chap. 6.

The stability analysis, however, can still be carried out, using the Von Neumann method, and the result show that the implicit scheme is unconditionally stable.

Problem: Show that the implicit scheme is first-order accurate in x and second-order accurate in y.

Show that the amplification factor for scheme (5.13) is given by

$$|g|^2 = \frac{1}{1 + 2\sigma^2(1 - \cos\alpha)} \le 1, \ \forall\alpha.$$

6. Integration of a Linear Parabolic Equation

6.1 Introduction

Parabolic PDEs represent diffusion phenomena, such as heat conduction in a slab aligned with the x-axis (2.8) and viscous effects spreading away from a wall. The numerical methods will be presented for the flow of an incompressible viscous fluid between two plates, distant h apart. Initially the fluid has a solid body motion with the plates translating parallel to the x-axis with velocity U. At time $t = 0$, the lower plate located at $y = 0$ is stopped, while the other plate keeps moving with velocity U. The equation governing the motion of the fluid is

$$\frac{\partial u}{\partial t} = \nu \frac{\partial^2 u}{\partial y^2}, \ t \geq 0, \ 0 \leq y \leq h. \tag{6.1}$$

$\nu \geq 0$ is the kinematic viscosity of the fluid.

The initial condition corresponds to

$$u(y, 0) = U, \ 0 \leq y \leq h, \tag{6.2}$$

and the boundary conditions are

$$\begin{cases} u(0, t) = 0 \\ u(h, t) = U \end{cases}, \ t > 0. \tag{6.3}$$

For large times, provided the dimensionless grouping $R_e = Uh/\nu$ is not too large, the flow reaches steady-state, i.e. $u(y, t) \to u_{ss}(y)$, which is independent of time. The solution is easily found to be $u_{ss}(y) = U(y/h)$, the so-called *Couette flow*. On the other hand, for very small times, the presence of the upper plate is not affecting the solution, i.e. the solution is the same as if $h = \infty$. In this case the *Rayleigh solution* is a good approximation

$$u(y, t) = \frac{2U}{\sqrt{\pi}} \int_0^{\frac{y}{2\sqrt{\nu t}}} e^{-\eta^2} d\eta.$$

6.2 Exact Solution

Equation (6.1) can be solved by the method of separation of variables. Assuming that $u(y,t) = Y(y)T(t)$, substitution into equation (6.1) yields

$$\frac{T'}{T} = \nu \frac{Y''}{Y} = -\lambda^2, \ \lambda = \text{const} \geq 0.$$

The choice of the minus sign in front of the constant is guided by the need for the solution to remain bounded as $t \to \infty$. The two ODEs can be integrated to give

$$u_\lambda(y,t) = \left(A_\lambda \sin \left(\frac{\lambda y}{\sqrt{\nu}} \right) + B_\lambda \cos \left(\frac{\lambda y}{\sqrt{\nu}} \right) \right) e^{-\lambda^2 t}.$$

At this point the parameter λ is arbitrary. Linear combinations of, differentiation and integration with respect to λ are allowed since the governing equation is linear. $\lambda = 0$ corresponds to steady-state, $u_0(y) = A_0 y + B_0 = u_{ss}(y)$.

Problem: Show that the steady-state equation corresponds to

$$u_{ss}(y) = \lim_{\lambda \to 0} \frac{\partial}{\partial \lambda} \left(\frac{U \sqrt{\nu}}{h} \sin \left(\frac{\lambda y}{\sqrt{\nu}} \right) e^{-\lambda^2 t} \right).$$

One requires that each $u_\lambda(y,t)$ satisfy the boundary conditions (6.3). Hence:

$$\begin{cases} u_0(0) = B_0 = 0 \\ u_0(h) = A_0 h = U, \Rightarrow A_0 = \frac{U}{h}, \end{cases}$$

$$\begin{cases} u_\lambda(0,t) = B_\lambda e^{-\lambda^2 t} = 0, \Rightarrow B_\lambda = 0 \\ u_\lambda(h,t) = A_\lambda \sin \frac{\lambda h}{\sqrt{\nu}} e^{-\lambda^2 t} = 0, \Rightarrow \frac{\lambda h}{\sqrt{\nu}} = k\pi, \ k = 1,2,... \end{cases}$$

The last condition defines a discrete spectrum for the *eigenvalues* λ_k

$$\lambda_k = \frac{k\pi \sqrt{\nu}}{h}, \ k = 1,2,... \tag{6.4}$$

The corresponding *eigenfunctions* are:

$$u_k(y,t) = A_k \sin \left(\frac{k\pi y}{h} \right) e^{-\left(k^2 \pi^2 \nu t / h^2\right)}.$$

The solution to (6.1)+(6.3) now reads

$$u(y,t) = U \frac{y}{h} + \sum_{k=1}^{\infty} A_k \sin \left(\frac{k\pi y}{h} \right) e^{-\left(k^2 \pi^2 \nu t / h^2\right)}.$$

Note that the summation corresponds to the discrete spectrum of eigenfunctions. The coefficients A_k are determined by the initial condition (6.2)

$$u(y,0) = U\frac{y}{h} + \sum_{k=1}^{\infty} A_k \sin\left(\frac{k\pi y}{h}\right) = U.$$

The A_k are the Fourier coefficients of the Fourier expansion of $U(1 - (y/h))$ as an odd series of period $2h$. The function is shown in Fig. 6.1. The coefficients are found to be $A_k = \frac{2U}{k\pi}$. Finally, the complete solution reads:

$$u(y,t) = U\frac{y}{h} + \frac{2U}{\pi}\sum_{k=1}^{\infty}\frac{1}{k}\sin\left(\frac{k\pi y}{h}\right)e^{-\left(k^2\pi^2\nu t/h^2\right)}.$$

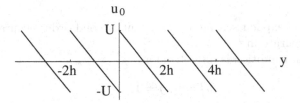

Fig. 6.1. Odd extension of the initial condition with period $2h$

Remark: The Fourier series is not absolutely convergent at $(0,0)$, because there is a discontinuity in the initial condition. However, for $t = \varepsilon$, as small as one wishes, the solution is continuous and differentiable to all orders.

6.3 A Simple Explicit Scheme

Let $y_j = (j-1)\Delta y$, $\Delta y = h/(jx - 1)$ and $t^n = n\Delta t$, represent the Cartesian mesh system. A simple explicit scheme is introduced as

$$\frac{u_j^{n+1} - u_j^n}{\Delta t} = \nu\frac{u_{j+1}^n - 2u_j^n + u_{j-1}^n}{\Delta y^2}, \quad j = 2, ..., jx - 1, \; n = 0, 1, ... \quad (6.5)$$

The mesh system and computational molecule are shown in Fig. 6.2.

6.3.1 Consistency and Accuracy

The TE for this scheme is

$$\epsilon_j^n = \frac{\partial u_j^n}{\partial t} + \frac{\Delta t}{2}\frac{\partial^2 u_j^n}{\partial t^2} + \frac{\Delta t^2}{3!}\frac{\partial^3 u_j^n}{\partial t^3} + O(\Delta t^3) - \nu\left(\frac{\partial^2 u_j^n}{\partial y^2} + \frac{\Delta y^2}{4!}\frac{\partial^4 u_j^n}{\partial y^4} + O(\Delta y^4)\right)$$

$$= O(\Delta t, \Delta y^2). \quad (6.6)$$

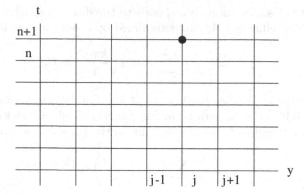

Fig. 6.2. Mesh system and computational molecule for scheme (6.5)

The simple explicit scheme is consistent, second-order accurate in y and first-order accurate in t.

The initial condition is discretized as

$$u_j^0 = U, \; j = 1, ..., jx,$$

and the boundary conditions

$$\begin{cases} u_1^{n+1} = 0 \\ u_{jx}^{n+1} = U \end{cases}, \; n = 0, 1, ... \; .$$

The initial and boundary conditions do not introduce discretization errors.

6.3.2 Numerical Implementation

Dimensionless variables are introduced. It is always a good idea to work with dimensionless quantities when solving a model from physics. Here we let $\tilde{y} = y/h$, $\tilde{u} = u/U$, $\tilde{t} = \nu t/h^2$, $\sigma = \nu(\Delta t/\Delta y^2)$. The equations with the dimensionless variables are identical to the dimensional equations in which one sets $\nu = U = h = 1$. The tilde will be dropped subsequently. In update form the scheme reads

$$\begin{cases} u_j^{n+1} = (1 - 2\sigma)u_j^n + \sigma(u_{j+1}^n + u_{j-1}^n), \; j = 2, ..., jx - 1, \; n = 0, 1, ... \\ u_j^0 = 1, \; j = 1, ...jx \\ u_1^{n+1} = 0, \; n = 0, 1, ... \\ u_{jx}^{n+1} = 1, \; n = 0, 1, ... \end{cases} \quad . \; (6.7)$$

The solution depends on a single parameter σ.

Problem: Solve (6.7) by hand with $jx = 5$ for $\sigma = \frac{1}{3}, \frac{1}{2}$, and $\sigma = 1$.

6.3.3 Stability

We recall the result obtained earlier that the stability condition for this scheme is $\sigma \leq \frac{1}{2}$, (2.11), or with the current notation

$$\Delta t \leq \frac{\Delta y^2}{2\nu}.$$

For a given Δy this condition imposes a restriction on the time step Δt that can be used to compute the solution. As noted earlier, as the mesh is refined in the y-direction, the time step decreases quadratically. The reason has to do with the requirement that in the limit, as $\Delta y \to 0$, the numerical domain of dependence should coincide with the domain of dependence of the PDE. As we have seen, the speed of propagation for the heat equation is infinite. As the mesh is refined, the slope of the numerical characteristics have an asymptotic behavior which is consistent with this result since $(dy/dt)_{N\pm} = \pm(\Delta y/\Delta t) = \pm O(1/\Delta y) \to \pm\infty$, as $\Delta y \to 0$. This is illustrated in Fig. 6.3, where the mesh has been divided by two in the y-direction and by four in the t-direction. On the Figure are also shown the numerical characteristics and the characteristics of the PDE.

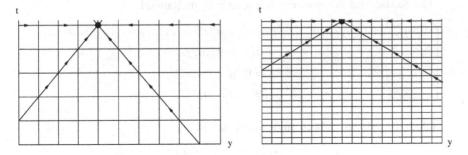

Fig. 6.3. Effect of mesh refinement on the domain of dependence

Problem: Study the stability of the DuFort-Frankel scheme for the heat equation (2.10)

$$\frac{u_i^{n+1} - u_i^{n-1}}{2\Delta t} = \alpha \frac{u_{i+1}^n - u_i^{n+1} - u_i^{n-1} + u_{i-1}^n}{\Delta x^2}.$$

In order to mimic more closely the physics of diffusion, and at the same time alleviate the restriction on the time step, various implicit schemes have been proposed.

6.4 Simple Implicit Method

The following scheme was proposed by Laasonen in 1949:

$$\frac{u_j^{n+1} - u_j^n}{\Delta t} = \nu \frac{u_{j+1}^{n+1} - 2u_j^{n+1} + u_{j-1}^{n+1}}{\Delta y^2}. \tag{6.8}$$

The computational molecule is displayed in Fig. 6.4.

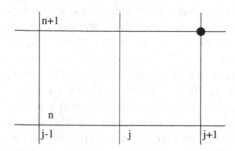

Fig. 6.4. Computational molecule for scheme (6.8)

The initial and boundary conditions are unchanged

$$\begin{cases} -\sigma u_{j-1}^{n+1} + (1 + 2\sigma)u_j^{n+1} - \sigma u_{j+1}^{n+1} = u_j^n, \ j = 2, ..., jx - 1, \ n = 0, 1, ... \\ u_j^0 = 1, \ j = 1, ..., jx \\ u_1^{n+1} = 0, \ n = 0, 1, ... \\ u_{jx}^{n+1} = 1, \ n = 0, 1, ... \end{cases}$$

$$\tag{6.9}$$

Changing Δt to $-\Delta t$ in (6.6) gives the TE

$$\epsilon_j^{n+1} = \frac{\partial u_j^{n+1}}{\partial t} - \frac{\Delta t}{2} \frac{\partial^2 u_j^{n+1}}{\partial t^2} + \frac{\Delta t^2}{3!} \frac{\partial^3 u_j^n}{\partial t^3} + O(\Delta t^3)$$

$$-\nu \left(\frac{\partial^2 u_j^{n+1}}{\partial y^2} + \frac{\Delta y^2}{4!} \frac{\partial^4 u_j^{n+1}}{\partial y^4} + O(\Delta y^4) \right) = O(\Delta t, \Delta y^2).$$

The Laasonen scheme is consistent, second-order accurate in space and first-order accurate in time.

Equation (6.8) cannot be solved pointwise as the simple explicit scheme because it contains three unknowns. A simultaneous solution for all the u_j^{n+1} must be performed.

The analysis of stability is carried out in the usual manner, with the complex wave mode $u_j^n = g^n e^{ij\beta}$ to give

$$g = \frac{1}{1 + 2\sigma(1 - \cos \beta)}.$$

Since $\sigma \geq 0 \Rightarrow |g| \leq 1$, the simple implicit scheme is unconditionally stable.

Problem: Solve problem (6.9) by hand for $\sigma = 1$ and $\sigma = \infty$, for $jx = 5$.

6.5 Combined Method A

The simple explicit and the simple implicit schemes are special cases of the general algorithm

$$\frac{u_j^{n+1} - u_j^n}{\Delta t} = (1-\theta)\nu \frac{u_{j+1}^n - 2u_j^n + u_{j-1}^n}{\Delta y^2} + \theta\nu \frac{u_{j+1}^{n+1} - 2u_j^{n+1} + u_{j-1}^{n+1}}{\Delta y^2}, \quad (6.10)$$

where θ is a constant $(0 \leq \theta \leq 1)$.

The simple explicit scheme corresponds to $\theta = 0$, the simple implicit scheme to $\theta = 1$. This combined method is second-order accurate in space and first-order accurate in time, except for special cases, such as:

i) $\theta = \frac{1}{2} \Rightarrow \epsilon_j^{n+\frac{1}{2}} = O(\Delta t^2, \Delta x^2)$. This is *the Crank-Nicolson method*. It is second-order accurate in x and t,

ii) $\theta = \frac{1}{2} - \frac{\Delta y^2}{12\nu\Delta t} \Rightarrow \epsilon_j^n = O(\Delta t^2, \Delta y^4)$,

iii) $\theta = \frac{1}{2} - \frac{\Delta y^2}{12\nu\Delta t}$ and $\frac{\Delta y^2}{\nu\Delta t} = \sqrt{20} \Rightarrow \epsilon_j^n = O(\Delta t^2, \Delta y^6)$.

Problem: Prove that the combined method (6.10) is unconditionally stable if $\frac{1}{2} \leq \theta \leq 1$ and stable when $0 \leq \theta \leq \frac{1}{2}$ only if $0 \leq \sigma \leq \frac{1}{2-4\theta}$.

6.6 Solution of a Linear System with Tridiagonal Matrix

Algebraic systems of equations obtained by discretization of PDEs by finite difference schemes often have a very simple structure, with few non-zero entries along certain diagonals. The case of tridiagonal matrices, as obtained for example with schemes (5.13) and (6.10) is important and justifies the presentation of the *Thomas algorithm* in this paragraph.

The Thomas algorithm (or *double sweep method*) is a particular case of the Gauss elimination algorithm pertinent to tridiagonal matrices.

Definition: A matrix $A = (a_{i,j})$ is said to be *tridiagonal* if $a_{i,j} = 0$ for $j \neq i - 1$, i or $i + 1$.

Consider the $n \times n$ system with tridiagonal matrix

$$T(p, q, r).x = s.$$

The elements of T are stored in vector arrays; q is the main diagonal, p the diagonal just below it, and r the diagonal just above it. x represents the unknown vector and s contains the right-hand sides

$$\begin{bmatrix} q_1 & r_1 & & & & \\ p_2 & q_2 & r_2 & & & \\ & & \cdot & \cdot & \cdot & \\ & & p_i & q_i & r_i & \\ & & & \cdot & \cdot & \cdot \\ & & & & \cdot & \cdot & \cdot \\ & & & & p_n & q_n \end{bmatrix} \cdot \begin{bmatrix} x_1 \\ x_2 \\ \cdot \\ x_i \\ \cdot \\ \cdot \\ x_n \end{bmatrix} = \begin{bmatrix} s_1 \\ s_2 \\ \cdot \\ s_i \\ \cdot \\ \cdot \\ s_n \end{bmatrix}.$$

In the direct sweep, the equations are combined to eliminate the lower diagonal and to transform the system in an upper triangular system. In this process the diagonal term is normalized to unity. If $q_1 \neq 0$ the first equation can written as

$$x_1 + \widehat{r}_1 x_2 = \widehat{s}_1, \ \widehat{r}_1 = \frac{r_1}{q_1}, \ \widehat{s}_1 = \frac{s_1}{q_1}.$$

Let's assume that the $i - 1$st equation can be written in a similar form. Then it can be combined with the i-th equation to give

$$
\begin{array}{ccccccc}
x_{i-1} + & \widehat{r}_{i-1} x_i & & = & \widehat{s}_{i-1} & | & -p_i \\
p_i x_{i-1} + & q_i x_i & + r_i x_{i+1} = & s_i & | & +1 \\
\hline
0 & (q_i - p_i \widehat{r}_{i-1}) x_i & + r_i x_{i+1} = s_i - p_i \widehat{s}_{i-1} & & & &
\end{array}
$$

Or, equivalently if $q_i - p_i \widehat{r}_{i-1} \neq 0$

$$x_i + \widehat{r}_i x_{i+1} = \widehat{s}_i, \ \widehat{r}_i = \frac{r_i}{q_i - p_i \widehat{r}_{i-1}}, \ \widehat{s}_i = \frac{s_i - p_i \widehat{s}_{i-1}}{q_i - p_i \widehat{r}_{i-1}}, \ i = 1, ..., n. \quad (6.11)$$

Since this is true for $i = 1$ by induction this will be true for all values of i. The recurrence formulae (6.11) are not strictly applicable for $i = 1$ and $i = n$, but it makes programming simpler if they can be used for all values of the index. Hence we define

$$\begin{cases} p_1 = 0, \ \widehat{r}_0 = 0, \ \widehat{s}_0 = 0 \\ r_n = 0 \end{cases}.$$

At the end of the direct sweep, the system has been transformed to $T(0, 1, \widehat{r}).x = \widehat{s}$, an upper triangular matrix with two diagonals

$$\begin{bmatrix} 1 & \widehat{r}_1 & & & & \\ & 1 & \widehat{r}_2 & & & \\ & & \cdot & \cdot & \cdot & \\ & & & 1 & \widehat{r}_i & \\ & & & & \cdot & \cdot & \cdot \\ & & & \cdot & \cdot & \cdot & 1 & \widehat{r}_{n-1} \\ & & & & & & 1 \end{bmatrix} \cdot \begin{bmatrix} x_1 \\ x_2 \\ \cdot \\ x_i \\ \cdot \\ x_{n-1} \\ x_n \end{bmatrix} = \begin{bmatrix} \widehat{s}_1 \\ \widehat{s}_2 \\ \cdot \\ \widehat{s}_i \\ \cdot \\ \widehat{s}_{n-1} \\ \widehat{s}_n \end{bmatrix}.$$

During the inverse sweep, the solution is obtained by tackling the equations in reverse order as

$$\begin{cases} x_n = \widehat{s}_n \\ x_{n-1} = \widehat{s}_{n-1} - \widehat{r}_{n-1} x_n \\ \phantom{x_{n-1}} \vdots \\ x_i = \widehat{s}_i - \widehat{r}_i x_{i+1}, \ i = n, n-1, ..., 1 \end{cases} \tag{6.12}$$

Remarks: The recurrence formulae (6.11)–(6.12) indicate that the modified values $\widehat{r}_i$, $\widehat{s}_i$ can be stored in the arrays r and s respectively and that the solution x can be overwritten in s, thus saving three arrays of length n.

When the end values x_1, x_n are known, e.g. from *Dirichlet boundary conditions*, it is possible to avoid the shifting of indices and solve an $n \times n$ system, instead of an $(n-2) \times (n-2)$ system, by adding the two trivial equations

$$\begin{cases} q_1 x_1 = s_1, \ q_1 = \dfrac{1}{\epsilon} \\ q_n x_n = s_n, \ q_n = \dfrac{1}{\epsilon} \end{cases}, \quad \epsilon = 1.0\,E - 10\,,$$

which will force the correct end values, even if r_1 and p_n have been defined elsewhere, provided ϵ is small enough for the diagonal terms to dominate the other coefficients in the same row.

In the Thomas algorithm the diagonal term is used as "pivot", normalized to one in the elimination procedure, and, for the solution to proceed to its term, it is required that its coefficient $q_i - p_i \widehat{r}_{i-1} \neq 0$ for all i. This is the case if the matrix is diagonally dominant, i.e.

$$|q_i| \geq |p_i| + |r_i|\,. \tag{6.13}$$

In the example (2.5), the coefficients of the tridiagonal matrix associated with the centered schemes (2.6) are

$$\begin{cases} p_i = \dfrac{\epsilon}{\Delta x^2} - \dfrac{1}{2\Delta x} \\ q_i = -2\dfrac{\epsilon}{\Delta x^2} \\ r_i = \dfrac{\epsilon}{\Delta x^2} + \dfrac{1}{2\Delta x} \end{cases},$$

and do not satisfy condition (6.13) when $\epsilon \leq \Delta x/2$. For small values of ϵ, it is advisable to use an advanced scheme for du/dx as

$$\epsilon \frac{u_{i+1} - 2u_i + u_{i-1}}{\Delta x^2} + \frac{u_{i+1} - u_i}{\Delta x} - 2\epsilon - 2x_i = 0. \tag{6.14}$$

This scheme has the property of reinforcing the diagonal term, so that

$$\begin{cases} p_i = \dfrac{\epsilon}{\Delta x^2} \\ q_i = -2\dfrac{\epsilon}{\Delta x^2} - \dfrac{1}{\Delta x} \\ r_i = \dfrac{\epsilon}{\Delta x^2} + \dfrac{1}{\Delta x} \end{cases},$$

and condition (6.13) is fulfilled. The solution is only first-order accurate, but insensitive to rounding-off errors. The results in Fig. 6.5 are for $\epsilon = 10^{-9}$ and $\Delta x = 0.02$. There are no oscillations in the numerical solution. Due to the advanced scheme, the error accumulates near $x = 0$ where a small jump is noticeable.

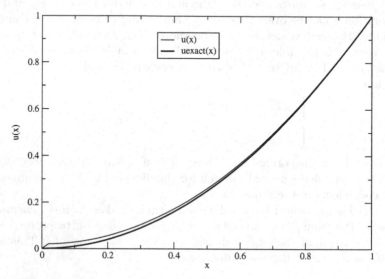

Fig. 6.5. Solution to (2.5) using (6.14) and a tridiagonal solver

7. Integration of a Linear Elliptic Equation

7.1 Introduction

Consider the following PDE, the so-called *Poisson's equation,* in the unit square:

$$\frac{\partial^2 u}{\partial x^2} + \frac{\partial^2 u}{\partial y^2} = f(x,y), \ (x,y) \in \Omega = [0,1] \times [0,1], \tag{7.1}$$

subject to Dirichlet boundary conditions

$$u(x,y) = g(x,y), \ (x,y) \in \partial\Omega. \tag{7.2}$$

$\partial\Omega$ represents the boundary of Ω. When $f = 0$ this is Laplace's equation.

Problem (7.1)–(7.2) models many physical phenomena, such as steady incompressible inviscid flows, equilibrium temperature in a square slab or static deformation of a membrane under distributed loads. In each situation, the independent variables are space-like and the problem is time independent.

In general, for arbitrary data g, it is not possible to find the exact solution using a simple technique. Complex variables and conformal mapping are too involved to be used, except in particular cases. As indicated before, we will make use of the source term f to construct exact solutions, in order to validate the numerical schemes. We will select low order polynomials to confirm the analysis of the truncation error. For example, $u(x,y) = x^2 - y^2$ is an exact solution with $f = 0$ and $g = x^2 - y^2$ on the boundary.

7.2 Numerical Scheme, Consistency, Accuracy

The mesh system is defined by

$$\begin{cases} x_i = (i-1)\Delta x, \ \Delta x = \frac{1}{ix-1} \\ y_j = (j-1)\Delta y, \ \Delta y = \frac{1}{jx-1} \end{cases},$$

and the numerical scheme by

$$\begin{cases} \dfrac{u_{i+1,j} - 2u_{i,j} + u_{i-1,j}}{\Delta x^2} + \dfrac{u_{i,j+1} - 2u_{i,j} + u_{i,j-1}}{\Delta y^2} = f_{i,j} \\ i = 2, ..., ix-1, \ j = 2, ..., jx-1 \end{cases}. \tag{7.3}$$

This is the classical five-points scheme.

The scheme is centered at point (i,j). This is the point chosen for the Taylor expansion. The TE reads

$$\epsilon_{i,j} = \frac{\partial^2 u_{i,j}}{\partial x^2} + \frac{\Delta x^2}{4!} \frac{\partial^4 u_{i,j}}{\partial x^4} + O(\Delta x^4) + \frac{\partial^2 u_{i,j}}{\partial y^2} + \frac{\Delta y^2}{4!} \frac{\partial^4 u_{i,j}}{\partial y^4} + O(\Delta y^4)$$
$$= O(\Delta x^2, \Delta y^2).$$

As expected, the scheme is consistent and second order accurate in x and in y. The computational molecule is shown in Fig. 7.1.

The boundary conditions are of Dirichlet type and read

$$\begin{cases} \begin{cases} u_{i,1} = g_{i,1} \\ u_{i,jx} = g_{i,jx}, \end{cases} , i = 1, ..., ix \\ \begin{cases} u_{1,j} = g_{1,j} \\ u_{ix,j} = g_{ix,j} \end{cases} , j = 1, ..., jx \end{cases} . \tag{7.4}$$

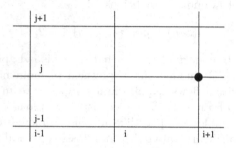

Fig. 7.1. Five-points scheme

7.3 Matrix Formulation; Direct Solution

The linear system of algebraic equations (7.3), together with the boundary conditions, can be recast in matrix form. We choose an ordering for the unknowns as a column vector in which the j-index is iterated upon before the i-index. The alternate choice is possible, and leads to a different matrix. let u be the vector

$$u = \begin{bmatrix} u_{1,1} \\ u_{1,2} \\ . \\ u_{1,jx} \\ . \\ u_{i,j} \\ . \\ u_{ix,jx} \end{bmatrix} .$$

Without detailing the boundary entries, the matrix associated with the linear system

$$A.x = b, \qquad (7.5)$$

has the following structure

$$A = \left[...0 \; \tfrac{1}{\Delta x^2} \; 0 \; ... \; 0 \; \tfrac{1}{\Delta y^2} \; -\tfrac{1}{h^2} \; \tfrac{1}{\Delta y^2} \; 0 \; ... \; 0 \; \tfrac{1}{\Delta x^2} \; 0... \right] .$$

For compactness we have introduced the notation $1/h^2 = 2((1/\Delta x^2) + (1/\Delta y^2))$. A is a pentadiagonal matrix. b is the vector containing the right-hand sides $f_{i,j}$ and the boundary conditions (7.4). A has at most five non-zero entries on each line. It is a large $(ix \times jx)^2$ sparse matrix with the entries along diagonals. This structure is a consequence of the structured mesh system used in finite differences.

The solution is $u = A^{-1}b$. The inverse matrix A^{-1} is expensive to compute and to store. In general it is a full matrix. Methods which do not require to compute A^{-1} are called *direct methods*.

Cramer's rule gives a way to compute the solution to (7.5), but the algorithm is immensely time consuming for $ix \times jx > 3$; the number of operations is proportional to $(ix \times jx + 1)!$, so it should not be used,

Gaussian elimination is a very useful and efficient tool, particularity for special cases, such as tridiagonal matrices, as we have seen. It requires approximately $(ix \times jx)^3$ operations.

Other more specific methods exist but must be considered on a case by case basis.

However, the most popular approach to solving large systems of algebraic equations is the *iterative method* that we turn to now.

7.4 Outlook of Iterative Methods

The basic principle of the iterative method is to construct a sequence of approximations u^n, $n = 0, 1, ...$ which satisfies equation (7.5) in the limit as $n \to \infty$. For sufficiently large n, the approximation can be considered good enough to terminate the iteration process.

We describe three well-established iterative methods:

− The Jacobi method
− The Gauss-Seidel method
− The over-relaxation method.

In our notation, u^n represents the vector state after n iterations, while u^0 represents an arbitrary initial vector. This reminds us of time-marching methods. Indeed, such a link exists between an iterative method and time dependent process, as we shall see later.

In the following we consider two-level methods, that is methods where the "new" values u^{n+1} depend only on the "old" values u^n. All such methods can be represented by

$$u^{n+1} - u^n = H(A.u^n - b), \tag{7.6}$$

where H is the *conditioning matrix* induced by the iterative method. H must be non-singular. In fact the iterative method defines the inverse matrix H^{-1} corresponding to $H^{-1}(u^{n+1} - u^n) = Au^n - b$.

Remarks: If $H = -A^{-1}$, i.e. if $H^{-1} = -A$, the exact solution is obtained in one single iteration, regardless of the initial guess. It is not possible in general to make such a selection, though. Intuitively, u^n will converge more rapidly to the solution u the closer H^{-1} is to $-A$.

The matrix A results from the discretization of the problem. A changes when the discretization scheme changes. The matrix H reflects the iterative method. H changes when the iterative method changes.

For linear problems, A, H and b do not depend on the index n. This is also the case for nonlinear problems, when close to the converged solution.

If we iterate the recurrence formula (7.6) and eliminate the intermediary levels, we find

$$u^{n+1} = (I + HA)^{n+1}u^0 - \sum_{m=0}^{n}(I + HA)^m Hb,$$

where I is the unit matrix. This result indicates that

i) if $u^{n+1} \to u$ as $n \to \infty$, then it must be true that $(I + HA)^{n+1} \to 0$, since u^0 is arbitrary,

ii) it must be also true that $\sum_{m=0}^{\infty}(I + HA)^m Hb \to A^{-1}b$, as $n \to \infty$.

Formally, the last expression operating on b represents the expansion of $-A^{-1}$ in series of matrices, an equivalent of the expansion in series of the scalar quantity:

$$-a^{-1} = \frac{-1}{a} = \frac{h}{1 - (1 + ha)} = \left(1 + (1 + ha) + (1 + ha)^2 + ...\right)h$$

$$= \sum_{m=0}^{\infty}(1 + ha)^m h.$$

This series converges for $|1 + ha| < 1$.

Let $|\lambda|_{\max}$ be the *spectral radius* of the matrix $I + HA$, i.e. $|\lambda|_{\max}$ is the modulus of the largest eigenvalue. Then, the necessary and sufficient condition for convergence is $|\lambda|_{\max} < 1$.

7.4.1 The Method of Jacobi

The method of Jacobi is best understood in the form

$$
\begin{cases}
\dfrac{u_{i+1,j}^{n} - 2u_{i,j}^{n+1} + u_{i-1,j}^{n}}{\Delta x^2} + \dfrac{u_{i,j+1}^{n} - 2u_{i,j}^{n+1} + u_{i,j-1}^{n}}{\Delta y^2} = f_{i,j} \ . \\
\qquad i = 2, ..., ix - 1, \ j = 2, ..., jx - 1.
\end{cases}
\tag{7.7}
$$

The new value for the unknown $u_{i,j}^{n+1}$ is obtained from the central term of the scheme, all the other unknowns being kept at their old values. Using the general formalism (7.6) we get

$$
\frac{1}{h^2}(u_{i,j}^{n+1} - u_{i,j}^{n}) = \frac{u_{i+1,j}^{n} - 2u_{i,j}^{n} + u_{i-1,j}^{n}}{\Delta x^2} + \frac{u_{i,j+1}^{n} - 2u_{i,j}^{n} + u_{i,j-1}^{n}}{\Delta y^2} - f_{i,j}.
\tag{7.8}
$$

In matrix form $u^{n+1} - u^n = h^2(Au^n - b)$. By identification, $H^{-1} = (1/h^2)I$, hence the conditioning matrix is $H = h^2 I$. The stability of the iterative procedure is governed by the modulus of the largest eigenvalue of $I + h^2 A$. Finding $|\lambda|_{\max}$ would allow us to conclude on the global stability, that is including the boundary conditions which are included in A. However, such a study is in general not practical if the matrix A has variable coefficients. The Von Neumann analysis is easier and the results are of great practical value.

Following this approach, and noting that the index n corresponds to the time-marching direction, while i, j correspond to space directions, let $u_{i,j}^n = g^n e^{ii\alpha} e^{ij\beta}$. Carrying this in equation (7.7), after some simplification one gets

$$
\frac{g}{h^2} = \frac{1}{\Delta x^2}(e^{i\alpha} + e^{-i\alpha}) + \frac{1}{\Delta y^2}(e^{i\beta} + e^{-i\beta}),
$$

that is

$$
g = \frac{\dfrac{\cos \alpha}{\Delta x^2} + \dfrac{\cos \beta}{\Delta y^2}}{\dfrac{1}{\Delta x^2} + \dfrac{1}{\Delta y^2}}.
$$

g is real and $-1 \leq g \leq 1$, $\forall \alpha, \beta$. The algorithm is stable.

The convergence speed is governed by the largest value of $|g|$. The mesh system can support wave modes that satisfy

$$
\begin{cases}
\alpha = k\pi\Delta x, \ k = 1, ..., ix - 1 \\
\beta = l\pi\Delta y, \ l = 1, ..., jx - 1
\end{cases}.
$$

Hence

$$
|g|_{\max} = 2h^2 \left(\frac{\cos \pi \Delta x}{\Delta x^2} + \frac{\cos \pi \Delta y}{\Delta y^2} \right) \cong 1 - 2\pi^2 h^2,
$$

for small values of the discretization steps. After N steps the amplitude of the error is reduced to $|g|_{max}^N$. The smaller $|g|_{max}$, the faster the convergence will be. For Jacobi's method $|g|_{max}$ is very close to one and $|g|_{max} \to 1$, as $h \to 0$, thus the convergence of Jacobi's method is expected to be very slow on fine meshes.

It is useful to define the *convergence rate* as $R = -\ln|g|_{max} \cong 2\pi^2 h^2$.

Remarks: Jacobi's method does not require us to follow any particular order in updating the points. However, it requires us to store temporarily the "new" values in an array u^{n+1} before they can be overwritten in the array u^n. If the domain is swept column by column, two extra arrays of length jx are needed for $u_{i-1,j}^{n+1}$, $u_{i,j}^{n+1}$, $j = 1, ..., jx$.

The interpretation of this method as an evolution process is straightforward from equation (7.8) by choosing, for example, the "time step" to be $\Delta\tau = h^2$. The Taylor expansion reads:

$$\frac{\partial u}{\partial \tau} = \frac{\partial^2 u}{\partial x^2} + \frac{\partial^2 u}{\partial y^2} + O(\Delta\tau, \Delta x^2, \Delta y^2).$$

This is the two-dimensional heat equation. The time step $\Delta\tau = h^2$ is the limit of the stability condition for the heat equation using an explicit centered five-points scheme in space and advanced scheme in time. Our study of the heat equation in Chap. 6 has shown that this type of scheme is very time consuming on fine mesh systems.

Problem: Solve by hand, using the method of Jacobi, Laplace's equation in the unit square $[0, 3] \times [0, 3]$ for a 4×4 mesh system, i.e. $x_i = (i-1)\Delta x$, $y_j = (j-1)\Delta y$, $\Delta x = \Delta y = 1$, and boundary conditions $g = (i-1)^2 - (j-1)^2$.

7.4.2 The Gauss–Seidel Method

The Gauss-Seidel method is more complex to analyze than Jacobi's method, yet in many ways, it is simpler to implement in a code, as it does not require any additional storage, besides that for the basic arrays.

The algorithm is as follows:

$$\frac{u_{i+1,j}^n - 2u_{i,j}^{n+1} + u_{i-1,j}^{n+1}}{\Delta x^2} + \frac{u_{i,j+1}^n - 2u_{i,j}^{n+1} + u_{i,j-1}^{n+1}}{\Delta y^2} = f_{i,j}. \qquad (7.9)$$

If one compares (7.7) and (7.9), one notices that $u_{i-1,j}$ and $u_{i,j-1}$ are new $(n+1)$ values in the scheme. However, it does not imply that the scheme is implicit and requires simultaneous solution of algebraic equations. In fact it refers to the fact that the grid points are swept with increasing values of the indices i and j and that these values are available from previously computed points. Hence, the only unknown in this equation is $u_{i,j}^{n+1}$. With the formalism above, this is

$$-\frac{1}{\Delta x^2}(u_{i-1,j}^{n+1} - u_{i-1,j}^n) - \frac{1}{\Delta y^2}(u_{i,j-1}^{n+1} - u_{i,j-1}^n) + \frac{1}{h^2}(u_{i,j}^{n+1} - u_{i,j}^n)$$

$$= \frac{u_{i+1,j}^n - 2u_{i,j}^n + u_{i-1,j}^n}{\Delta x^2} + \frac{u_{i,j+1}^n - 2u_{i,j}^n + u_{i,j-1}^n}{\Delta y^2} - f_{i,j}.$$

By identification one finds

$$H^{-1} = \left[...0 - \tfrac{1}{\Delta x^2}\, 0\, ...\, 0 - \tfrac{1}{\Delta y^2}\, \tfrac{1}{h^2}\, 0\, 0... \right].$$

H^{-1} is a lower triangular matrix and its coefficients are the opposite of the corresponding coefficients in A.

The method of Gauss-Seidel is a particular case of the over-relaxation method and will be studied in this more general framework.

Problem: Solve by hand, using the method of Gauss-Seidel, the Laplace's equation in the unit square $[0,3] \times [0,3]$ for a 4×4 mesh system, i.e. $x_i = (i-1)\Delta x$, $y_j = (j-1)\Delta y$, $\Delta x = \Delta y = 1$, and boundary conditions $g = (i-1)^2 - (j-1)^2$.

7.4.3 The Successive Over-Relaxation Method (SOR)

The method is in two steps:

i) First compute a provisional value $\tilde{u}_{i,j}$ using the Gauss–Seidel formula

$$\frac{u_{i+1,j}^n - 2\tilde{u}_{i,j} + u_{i-1,j}^{n+1}}{\Delta x^2} + \frac{u_{i,j+1}^n - 2\tilde{u}_{i,j} + u_{i,j-1}^{n+1}}{\Delta y^2} = f_{i,j}. \tag{7.10}$$

ii) Then extrapolate a new value using the relation

$$u_{i,j}^{n+1} = u_{i,j}^n + \omega(\tilde{u}_{i,j} - u_{i,j}^n), \tag{7.11}$$

where ω is the *relaxation factor*. When $\omega > 1$, this is called *over-relaxation*. In certain situations, e.g. nonlinear equations, it is necessary to have $0 < \omega < 1$ to maintain the stability of the process; it is called then *under-relaxation*. When $\omega = 1$, this is the Gauss-Seidel Method.

The two steps can be combined into a single formula by elimination of the provisional value

$$\tilde{u}_{i,j} = u_{i,j}^n + \frac{1}{\omega}(u_{i,j}^{n+1} - u_{i,j}^n).$$

The formal expression of the iterative process is

$$-\frac{1}{\Delta x^2}(u_{i-1,j}^{n+1} - u_{i-1,j}^n) - \frac{1}{\Delta y^2}(u_{i,j-1}^{n+1} - u_{i,j-1}^n) + \frac{1}{\omega h^2}(u_{i,j}^{n+1} - u_{i,j}^n)$$

$$= \frac{u_{i+1,j}^n - 2u_{i,j}^n + u_{i-1,j}^n}{\Delta x^2} + \frac{u_{i,j+1}^n - 2u_{i,j}^n + u_{i,j-1}^n}{\Delta y^2} - f_{i,j}. \tag{7.12}$$

From this we can identify the inverse of the conditioning matrix to be

$$H^{-1} = \left[...0 - \tfrac{1}{\Delta x^2}\ 0\ ...\ 0 - \tfrac{1}{\Delta y^2}\ \tfrac{1}{\omega h^2}\ 0\ 0... \right].$$

The Von Neumann analysis is carried out with $f = 0$ and the complex wave mode $u_{i,j}^n = g^n e^{ii\alpha} e^{ij\beta}$. After some algebra one arrives at

$$\left(\frac{1 - \cos\alpha}{\Delta x^2} + \frac{1 - \cos\beta}{\Delta y^2} + \frac{2 - \omega}{2\omega h^2} + i \left(\frac{\sin\alpha}{\Delta x^2} + \frac{\sin\beta}{\Delta y^2} \right) \right) g =$$
$$- \left(\frac{1 - \cos\alpha}{\Delta x^2} + \frac{1 - \cos\beta}{\Delta y^2} - \frac{2 - \omega}{2\omega h^2} \right) + i \left(\frac{\sin\alpha}{\Delta x^2} + \frac{\sin\beta}{\Delta y^2} \right). \tag{7.13}$$

The over-relaxation method is stable for $0 < \omega < 2$, because $|g| \le 1$, as can be seen from (7.13). For each value of ω, there is a maximum value of $|g|$ corresponding to $\alpha = \pi\Delta x$, $\beta = \pi\Delta y$. Let $g(\omega)$ be that value; it is of the form $g(\omega) = (-a_1 + ib)/(a_2 + ib) =$, where a_1, a_2 and b are real numbers. As ω varies, it has been shown that there exists a value ω_{opt} which minimizes the amplification factor. It has also been shown that as $h \to 0$, $\omega_{opt} \to 2$. The curve $g(\omega)$ is sketched in Fig. 7.2.

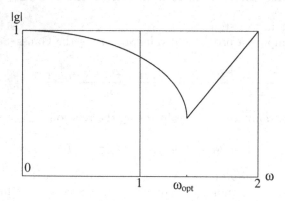

Fig. 7.2. Amplification factor versus ω

Let $\omega = 2 - ch$ as $h \to 0$, where $c = $ const. The leading terms of $g(\omega)$ is found to be

$$|g(\omega_{opt})| = 1 - \frac{\pi^2 ch}{\pi^2 + \frac{c^2}{4}} + O(h^2)\ h \to 0.$$

The optimum is reached for $c = 2\pi$, that is $|g(\omega_{opt})| = 1 - \pi h + O(h^2)$. The relaxation method is converging faster than Jacobi's method on fine meshes. The convergence rate is $R = \pi h$, that is one order of magnitude faster than Jacobi's.

One could show that the Gauss-Seidel method is only twice as fast as the Jacobi method.

Remark: Equation (7.12) can be interpreted as the discretization of an evolution equation, upon expanding in Taylor series about (i, j, n):

$$\Delta\tau \frac{2 - \omega}{2\omega h^2} \frac{\partial u_{i,j}^n}{\partial \tau} + \frac{\Delta\tau}{\Delta x} \frac{\partial^2 u_{i,j}^n}{\partial \tau \partial x} + \frac{\Delta\tau}{\Delta y} \frac{\partial^2 u_{i,j}^n}{\partial \tau \partial y} + O\left((2 - \omega)\frac{\Delta\tau^2}{h^2}, \frac{\Delta\tau^2}{\Delta x}, \frac{\Delta\tau^2}{\Delta y}\right)$$

$$= \frac{\partial^2 u_{i,j^n}}{\partial x^2} + \frac{\partial^2 u_{i,j}}{\partial y^2} - f_{i,j} + O(\Delta x^2, \Delta y^2)$$

i) If $\omega = 1$ or $\omega \neq 2$, $h \to 0$, then choosing $\Delta\tau = 2\omega h^2/(2 - \omega) = O(h^2)$ yields the heat equation with its slow convergence on fine meshes (same as for Jacobi's method).

ii) If $\omega = 2 - 2\pi h$, $h \to 0$, then choosing $\Delta\tau = 2\omega h^2/(2 - \omega) = O(h)$ yields a different equation in the limit, i.e.

$$\frac{\partial u}{\partial \tau} + a\frac{\partial^2 u}{\partial \tau \partial x} + b\frac{\partial^2 u}{\partial \tau \partial y} = \frac{\partial^2 u}{\partial x^2} + \frac{\partial^2 u}{\partial y^2} - f.$$

This equation is hyperbolic and converges faster than the heat equation on fine grids.

As an example, Laplace's equation is solved on the unit square. The exact solution is $u(x, y) = x^2 - y^2$ ($f = 0$). The mesh is 41×41, i.e. $\Delta x = \Delta y = 0.025$. The initial condition is $u_{i,j}^0 = 0$, except on the boundaries where the exact value is imposed. The residual Res is defined as

$$Res(n) = \frac{u_{i+1,j}^n - 2u_{i,j}^n + u_{i-1,j}^n}{\Delta x^2} + \frac{u_{i,j+1}^n - 2u_{i,j}^n + u_{i,j-1}^n}{\Delta y^2} - f_{i,j}.$$

For comparison purposes, a normalized residual is used in the plot as $R = Res(n)/Res(0)$. The results are displayed in Fig. 7.3, for the method of Jacobi, Gauss-Seidel and over-relaxation ($\omega = 1.75$). Although the mesh is not very fine, the theoretical results are well recovered. In particular the convergence rate for Jacobi's method is found from the graph to be $R = 0.0032 \cong 2\pi^2 h^2$, the Gauss-Seidel method converges twice as fast as Jacobi's and the optimized relaxation method has a convergence rate $R = 0.04 \cong \pi h$. Note that the optimum value of ω has been estimated by trial and error.

The converged solution is the exact solution to machine accuracy.

7.5 Other Iterative Methods

The successive over-relaxation method (SOR) studied above was found superior to Jacobi's and Gauss-Seidel methods thus far. Block iterative methods can be superior to SOR in many cases, in particular when the mesh is stretched in the direction where the scheme is implicit. In contrast to the latter, the former singles out subgroups of unknowns which are solved simultaneously

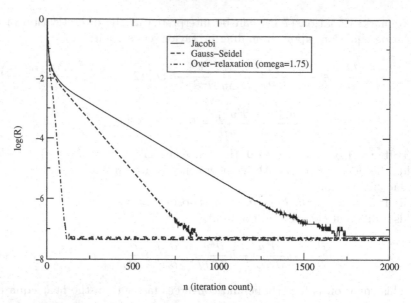

Fig. 7.3. Comparison of residual evolution with iterations

using an efficient procedure such as the Thomas algorithm. For example, with reference to the model problem, choosing columns of unknowns to be solved with the Thomas algorithm is known as the *successive line over-relaxation method* (SLOR). Another class of iterative methods is the alternating direction implicit method or ADI in short. We briefly study these now.

7.5.1 The SLOR Method

This method has two steps:
  i) Compute the provisional values $\widetilde{u}$ from

$$\frac{u^n_{i+1,j} - 2\widetilde{u}_{i,j} + u^{n+1}_{i-1,j}}{\Delta x^2} + \frac{\widetilde{u}_{i,j+1} - 2\widetilde{u}_{i,j} + \widetilde{u}_{i,j-1}}{\Delta y^2} = f_{i,j}. \tag{7.14}$$

  ii) Get the final value as

$$u^{n+1}_{i,j} = u^n_{i,j} + \omega(\widetilde{u}_{i,j} - u^n_{i,j}).$$

The scheme can be rewritten as

$$-\frac{1}{\Delta x^2}(u^{n+1}_{i-1,j} - u^n_{i-1,j}) - \frac{1}{\omega \Delta y^2}(u^{n+1}_{i,j-1} - u^n_{i,j-1})$$

$$+\frac{1}{\omega h^2}(u^{n+1}_{i,j} - u^n_{i,j}) - \frac{1}{\omega \Delta y^2}(u^{n+1}_{i,j+1} - u^n_{i,j+1})$$

$$= \frac{u^n_{i+1,j} - 2u^n_{i,j} + u^n_{i-1,j}}{\Delta x^2} + \frac{u^n_{i,j+1} - 2u^n_{i,j} + u^n_{i,j-1}}{\Delta y^2} - f_{i,j}.$$

Note that the first term in the left-hand side is known from the solution of the previous column. The current unknowns form a tridiagonal system that is solved efficiently using the Thomas algorithm. The matrix is diagonally dominant. The relaxation factor is constrained between zero and two, $0 < \omega < 2$.

Although this method requires more operations per sweep than SOR procedures, it is in general less expensive because it requires less sweeps for convergence. This is due to the fact that the boundary conditions at both ends of a column are felt immediately within the domain, thanks to the implicitness of the scheme along a column. The explicitness remains in the i-direction, and the sweeps allow for the boundary condition at $i = ix$ to propagate towards $i = 1$. It takes ix sweeps for that boundary condition to be felt in the whole domain.

As an example, the SOR and SLOR methods are compared for the solution of Laplace's equation on the unit square. The mesh is 101×101 and the relaxation factors are $\omega_{SOR} = 1.85$, $\omega_{SLOR} = 1.9$. The convergence curves are plotted in Fig. 7.4. There is not much difference between the two methods on this mesh system. One clearly sees the break in the convergence curve for the SLOR when the influence of the boundary condition at $i = ix$ reaches the boundary $i = 1$ after 100 sweeps. Also visible is the higher roundoff error level for the converged solution using SLOR of 2.10^{-7} versus 5.10^{-8} for SOR. This is due to some loss of accuracy in the tridiagonal solver algorithm.

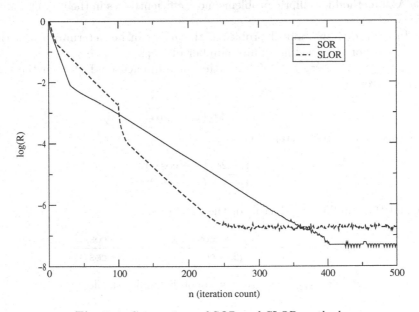

Fig. 7.4. Comparison of SOR and SLOR methods

7.5.2 ADI Methods

Schemes patterned after the ADI procedure for the two-dimensional heat equation have been applied to more complex problems including transonic potential flows, Euler and Navier-Stokes simulations. When the interest resides in the steady-state solution, the size of the time-step can be chosen with a view to accelerate convergence of the iterative process.

The two-step Peaceman-Rachford's ADI scheme for Poisson's equation reads:

i) step 1

$$u_{i,j}^{n+\frac{1}{2}} = u_{i,j}^n + \rho^n \left(\frac{u_{i+1,j}^{n+\frac{1}{2}} - 2u_{i,j}^{n+\frac{1}{2}} + u_{i-1,j}^{n+\frac{1}{2}}}{\Delta x^2} + \frac{u_{i,j+1}^n - 2u_{i,j}^n + u_{i,j-1}^n}{\Delta y^2} - f_{i,j} \right),$$

ii) step 2

$$u_{i,j}^{n+1} = u_{i,j}^{n+\frac{1}{2}} + \rho^n \left(\frac{u_{i+1,j}^{n+\frac{1}{2}} - 2u_{i,j}^{n+\frac{1}{2}} + u_{i-1,j}^{n+\frac{1}{2}}}{\Delta x^2} + \frac{u_{i,j+1}^{n+1} - 2u_{i,j}^{n+1} + u_{i,j-1}^{n+1}}{\Delta y^2} - f_{i,j} \right).$$

The ρ^n are known as *iteration parameters*. The Peaceman-Rachford iterative procedure for solving Laplace's equation on a square is convergent for any fixed value of ρ^n. On the other hand, for maximum computational efficiency, the iteration parameters should be cycled with n. The key to using the ADI method for elliptic problems most efficiently lies in the proper choice of the ρ^n's.

In certain simple model problems, the ρ^n's can be determined and the exact solution obtained in a finite number of steps.

Let $\sigma_x = \rho^n / \Delta x^2$, $\sigma_y = \rho^n / \Delta y^2$, the the amplification factor for the first step is given by

$$g_1 = \frac{1 - 2\sigma_y(1 - \cos\beta)}{1 + 2\sigma_x(1 - \cos\alpha)},$$

whereas for the second step it is

$$g_2 = \frac{1 - 2\sigma_x(1 - \cos\alpha)}{1 + 2\sigma_y(1 - \cos\beta)},$$

so that the amplification factor for the two steps is

$$g = g_1 g_2 = \frac{1 - 2\sigma_y(1 - \cos\beta)}{1 + 2\sigma_x(1 - \cos\alpha)} \frac{1 - 2\sigma_x(1 - \cos\alpha)}{1 + 2\sigma_y(1 - \cos\beta)}.$$

$|g| \le 1$, $\forall \alpha, \beta$. The ADI procedure is unconditionally stable.

8. Finite Difference Scheme
for a Convection-Diffusion Equation

8.1 Introduction

First-order model equations for convection have been studied in previous chapters, i.e. the linear convection equation (2.12) and the Burgers' equation (inviscid) (4.9). The heat equation (2.8) is a model for diffusion. When both phenomena are present, as for example in boundary-layer flows, the PDE must contain first as well as second partial derivatives in space. Such a model has been derived by Burgers and is known as the viscous Burgers' equation or simply Burgers' equation:

$$\frac{\partial u}{\partial t} + u\frac{\partial u}{\partial x} = \nu\frac{\partial^2 u}{\partial x^2}. \tag{8.1}$$

This equation was derived to study the internal structure of a weak normal shock in the near-sonic regime. $u(x,t)$ represents the perturbation from a uniform sonic stream ($u = 0$ corresponds to sonic condition).

Initial and boundary conditions must be added to complete the problem.

When convection and diffusion balance each other, a steady-state is reached. An exact steady solution to (8.1) that satisfies the following BC's

$$\begin{cases} u(0,t) = u_0 \\ u(L,t) = 0, \end{cases}$$

is given by

$$u(x) = u_0\overline{u}\frac{1 - e^{-\overline{u}Re_L(1-\frac{x}{L})}}{1 + e^{-\overline{u}Re_L(1-\frac{x}{L})}},$$

where $\overline{u}$ satisfies

$$\frac{\overline{u} - 1}{\overline{u} + 1} = e^{-\overline{u}Re_L}.$$

In turn Re_L is the *Reynolds number*, representing the ratio of convection to diffusion, $Re_L = \frac{u_0 L}{\nu}$.

Two limiting cases are important:

i) $Re_L \to 0$, in which case $\overline{u} \cong \sqrt{\frac{2}{Re_L}}$, and the highly viscous flow limit is $u(x) = u_0(1 - \frac{x}{L})$

ii) $Re_L \to \infty$, thus $\overline{u} \to 1$; the singular inviscid limit is

$$\begin{cases} u(x) = u_0, \ 0 \le x < L \\ \qquad u(L) = 0. \end{cases}$$

This is depicted in Fig. 8.1. This represents half of the shock structure. The remaining part is the mirror reflection through the point $(L,0)$ and extends to $x = 2$ where $u = -u_0$.

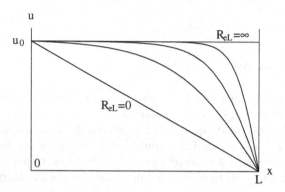

Fig. 8.1. Steady-state solution of the convection-diffusion equation

A somewhat simpler model is often used, which corresponds to the linearized Burgers' equation:

$$\frac{\partial u}{\partial t} + c\frac{\partial u}{\partial x} = \nu\frac{\partial^2 u}{\partial x^2}. \tag{8.2}$$

In this model the convection speed is constant and equal to c in the first half of the shock structure and $-c$ in the second half. The corresponding exact steady-state solution is

$$u(x) = u_0 \frac{1 - e^{-R_{eL}(1-\frac{x}{L})}}{1 - e^{-R_{eL}}},$$

where $R_{eL} = \frac{cL}{\nu}$.

8.2 FTCS Method

Roach (1972) (Reference [3]) has given the name *Forward-in-Time and Centered-in-Space method* to the scheme obtained by applying forward in time and centered in space finite difference operators to the linearized Burgers' equation (8.2). The scheme reads

$$\frac{u_i^{n+1} - u_i^n}{\Delta t} + c\frac{u_{i+1}^n - u_{i-1}^n}{2\Delta x} = \nu\frac{u_{i+1}^n - 2u_i^n + u_{i-1}^n}{\Delta x^2}. \tag{8.3}$$

The analysis of the TE is straightforward, expanding about point (i, n)

$$\epsilon_i^n = \frac{\partial u_i^n}{\partial t} + \frac{\Delta t}{2}\frac{\partial^2 u_i^n}{\partial t^2} + O(\Delta t^2) + c\left(\frac{\partial u_i^n}{\partial x} + \frac{\Delta x^2}{3!}\frac{\partial^3 u_i^n}{\partial x^3} + O(\Delta x^4)\right)$$

$$-\nu\left(\frac{\partial^2 u_i^n}{\partial x^2} + \frac{\Delta x^2}{4!}\frac{\partial^4 u_i^n}{\partial x^4} + O(\Delta x^4)\right) = O(\Delta t, \Delta x^2).$$

The FTCS scheme is consistent, first-order accurate in time and second-order accurate in space.

For the study of stability, let $\sigma = (c\Delta t/\Delta x)$, $r = (\nu\Delta t/\Delta x^2)$ and the complex solution be $u_i^n = g^n e^{\underline{i} i \alpha}$. One finds

$$g = 1 - 2r(1 - \cos\alpha) - \underline{i}\sigma\sin\alpha,$$

that is

$$|g|^2 = 1 + \left(8r^2 - 4r + \left(\sigma^2 - 4r^2\right)(1 + \cos\alpha)\right)(1 - \cos\alpha).$$

The scheme will be stable if $|g|^2 \le 1$, $\forall\alpha$.

i) If $\sigma^2 - 4r^2 \le 0$, the worst case corresponds to $\cos\alpha = -1$, and the condition reduces to $2r \le 1$,

ii) If $\sigma^2 - 4r^2 \ge 0$, the worst case corresponds to $\cos\alpha = 1$, and the condition reduces to $\sigma^2 \le 2r$.

The two cases are accounted for by

$$\sigma^2 \le 2r \le 1. \tag{8.4}$$

If one defines the cell Reynolds number $Re_{\Delta x} = \sigma/r$, the stability condition requires that the cell Reynolds number be restricted to

$$Re_{\Delta x} \le \frac{2}{\sigma}.$$

An important characteristic of finite difference schemes used for solving Burgers' equation is whether they produce oscillations (wiggles) in the solution. Obviously, it is desirable to have a solution that is free of oscillations. The FTCS method will produce oscillations in the solution to (8.2) when the cell Reynolds number is in the range

$$2 \le Re_{\Delta x} \le \frac{2}{\sigma}.$$

The oscillations occur because a monotone solution cannot produce, on a given mesh, a steeper gradient and curvature than that of the following data

$$\begin{cases} u_i = u_0, \ i = 1, ..., ix - 1 \\ \qquad u_{ix} = 0. \end{cases} \tag{8.5}$$

Only a solution with overshoots and undershoots will be able to produce higher values for $\partial u/\partial x$ and $\partial^2 u/\partial x^2$ near $x = L$.

The two stability conditions (8.4) can be expressed as

$$\begin{cases} 2r \leq 1 \implies \Delta t_v \leq \dfrac{\Delta x^2}{2\nu} \\[3mm] \sigma^2 \leq 2r \implies \Delta t_R \leq \dfrac{2\nu}{c^2}. \end{cases}$$

The first condition is the same as for the heat equation. The second condition states that the cell Reynolds number must be less than $2/\sigma$.

Results for the FTCS scheme are presented in Fig. 8.2 for $ix = 21$, $L = 1$, $c = 1$, and three values of the viscosity $\nu = 0.05$, 0.025 and 0.01.

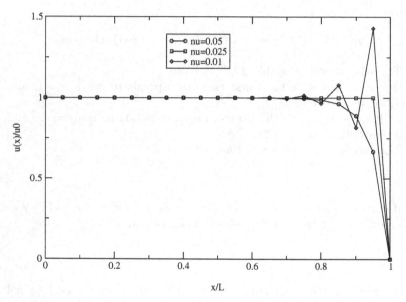

Fig. 8.2. Steady-state solution for the FTCS scheme for $\nu = 0.05$, 0.025 and 0.01

For large value of viscosity, the time step limitation is due to viscous effects with $\Delta t \leq \Delta t_v < \Delta t_R$, and for $\nu = 0.05$, the cell Reynolds number is $Re_{\Delta x} = 1$. No oscillations occur in this case.

The intermediate value of viscosity $\nu = 0.025$ is such that the two time steps are the same, i.e. $\Delta t \leq \Delta t_v = \Delta t_R$, and the cell Reynolds number is $Re_{\Delta x} = 2$. The solution corresponds to the data (8.5).

As the viscosity is further decreased, the stability condition is imposed by the cell Reynolds number, i.e. $\Delta t \leq \Delta t_R < \Delta t_v$. For $\nu = 0.01$ the cell Reynolds number is $Re_{\Delta x} = 5$. The oscillations are quite large.

8.3 The Box and Modified Box Methods

Keller introduced in 1970 a box method for parabolic PDEs. The differencing
is implicit, formally second-order accurate, and appears to differ from other
second-order accurate implicit procedures in that the formulation applies
equally to grids spacing that are arbitrary. Second derivatives are replaced
by first derivatives through the addition of new dependent variables and new
equations.

Equation (8.2) is written as a system

$$\begin{cases} \dfrac{\partial u}{\partial t} + c\dfrac{\partial u}{\partial x} = \nu\dfrac{\partial v}{\partial x} \\[2mm] \dfrac{\partial u}{\partial x} = v \end{cases}.$$

The first-order equation is discretized within each element, or box, using
the four nodal values from each corner as

$$\begin{cases} \dfrac{1}{2}\left(\dfrac{u_{i-1}^{n+1} - u_{i-1}^{n}}{\Delta t} + \dfrac{u_{i}^{n+1} - u_{i}^{n}}{\Delta t} \right) + \dfrac{c}{2}\left(\dfrac{u_{i}^{n+1} - u_{i-1}^{n+1}}{\Delta x} + \dfrac{u_{i}^{n} - u_{i-1}^{n}}{\Delta x} \right) \\[3mm] \qquad = \dfrac{\nu}{2}\left(\dfrac{v_{i}^{n+1} - v_{i-1}^{n+1}}{\Delta x} + \dfrac{v_{i}^{n} - v_{i-1}^{n}}{\Delta x} \right) \\[3mm] \dfrac{u_{i}^{n+1} - u_{i-1}^{n+1}}{\Delta x} = \dfrac{1}{2}\left(v_{i-1}^{n+1} + v_{i}^{n+1} \right). \end{cases}$$

This leads to a block tridiagonal system that can be solved by a block
elimination procedure. However, for this simple model equation, the variable
v can be eliminated altogether by combining adjacent equations. The result,
with the notation $\delta u_i = u_i^{n+1} - u_i^n$, is the so-called *delta-form* of the FDE

$$\left(\frac{1}{4\Delta t} - \frac{c}{4\Delta x} - \frac{\nu}{2\Delta x^2} \right) \delta u_{i-1} + \left(\frac{1}{2\Delta t} + \frac{\nu}{\Delta x^2} \right) \delta u_i$$

$$+ \left(\frac{1}{4\Delta t} + \frac{c}{4\Delta x} - \frac{\nu}{2\Delta x^2} \right) \delta u_{i+1}$$

$$= -c\frac{u_{i+1}^n - u_{i-1}^n}{2\Delta x} + \nu\frac{u_{i+1}^n - 2u_i^n + u_{i-1}^n}{\Delta x^2}.$$

This is different from the Crank-Nicolson scheme in the coefficients of the
time derivative terms. This scheme is consistent and second-order accurate in
space and time on uniform meshes. A conceptual advantage of schemes, such
as this one, based on the box molecule, is that formal second-order accuracy
is maintained, even when the mesh is non-uniform.

8.4 A Mixed-Type Scheme

A mixed-type scheme will be discussed in Chap. 10 for the inviscid Burgers' equation. It features four discretization operators which are selected according to the local orientation of the characteristics at a point, called a supersonic, shock, sonic or subsonic point. Based on the inequalities or *switches*, the four cases are:

i) $u_{i-\frac{1}{2}}^n > 0$, $u_{i+\frac{1}{2}}^n > 0$ (supersonic point):

$$\frac{u_i^{n+1} - u_i^n}{\Delta t} + u_{i-\frac{1}{2}}^n \frac{u_i^n - u_{i-1}^n}{\Delta x} = 0,$$

ii) $u_{i-\frac{1}{2}}^n > 0$, $u_{i+\frac{1}{2}}^n < 0$ (shock point):

$$\frac{u_i^{n+1} - u_i^n}{\Delta t} + u_{i-\frac{1}{2}}^n \frac{u_i^n - u_{i-1}^n}{\Delta x} + u_{i+\frac{1}{2}}^n \frac{u_{i+1}^n - u_i^n}{\Delta x} = 0,$$

iii) $u_{i-\frac{1}{2}}^n < 0$, $u_{i+\frac{1}{2}}^n > 0$ (sonic point):

$$\frac{u_i^{n+1} - u_i^n}{\Delta t} + u_i^{n+1} \frac{u_{i+1}^n - u_{i-1}^n}{2\Delta x} = 0,$$

iv) $u_{i-\frac{1}{2}}^n < 0$, $u_{i+\frac{1}{2}}^n < 0$ (subsonic point):

$$\frac{u_i^{n+1} - u_i^n}{\Delta t} + u_{i+\frac{1}{2}}^n \frac{u_{i+1}^n - u_i^n}{\Delta x} = 0,$$

where the mid-values (switches) correspond to averages as

$$u_{i-\frac{1}{2}} = \frac{u_{i-1} + u_i}{2}.$$

This scheme can be extended easily to the viscous Burgers' equation by adding the second derivative term in the right-hand side, discretized with a centered scheme as in (8.3). The benefit of this approach is that, since the inviscid scheme does not exhibit wiggles, the same will be true of the scheme with the viscous term, even in the limit as $\nu \to 0$.

Problem: Study the consistency and accuracy of the supersonic point with the viscous term in the right-hand side

i) $u_{i-\frac{1}{2}}^n > 0$, $u_{i+\frac{1}{2}}^n > 0$ (supersonic point):

$$\frac{u_i^{n+1} - u_i^n}{\Delta t} + u_{i-\frac{1}{2}}^n \frac{u_i^n - u_{i-1}^n}{\Delta x} = \nu \frac{u_{i+1}^n - 2u_i^n + u_{i-1}^n}{\Delta x^2}.$$

Study the stability condition in terms of $\sigma = u_{i-\frac{1}{2}}^n(\Delta t/\Delta x)$ and $r = \nu(\Delta t/\Delta x^2)$.

9. The Method of Murman and Cole

9.1 Introduction

At the end of the sixties, transonic flows solutions were attempted with the computer resources of the time, in order to help better understand and design transport aircraft in a regime where shock waves are present and have a major influence on the global performance in cruise. Most work was based on the solution of the unsteady Euler equations using some of the classical schemes introduced in Chap. 5, but the convergence was painstakingly slow. J.D. Cole and E.M. Murman met, while on stay at Boeing, and decided to look for an alternative approach to the problem, in which the steady-state equation is tackled directly. They chose as a model the transonic small disturbance equation (TSD), a familiar equation for J.D. Cole in his study of transonic similarity (Reference [4]).

The starting point was to remark that efficient methods existed for solving the TSD equation, in subsonic flow regime using SLOR, the equation being of elliptic type, and in the supersonic regime using marching techniques since the equation is hyperbolic in the flow direction.

Murman and Cole successfully combined both procedures in a single method, achieving unprecedented efficiency in the computation of mixed-type flows with shock waves.

9.2 The Model Problem

Consider two-dimensional steady inviscid compressible flow past a profile. Such a flow is governed by the Euler equations. A simpler model can be derived when assuming that the obstacle is a thin profile, thereby disturbing only slightly the flow from its uniform state. The equations of conservation of mass, momentum and energy are

$$\begin{cases} \dfrac{\partial \rho u}{\partial x} + \dfrac{\partial \rho v}{\partial y} = 0 \\[2mm] \dfrac{\partial \rho u^2 + p}{\partial x} + \dfrac{\partial \rho uv}{\partial y} = 0 \\[2mm] \dfrac{\partial \rho uv}{\partial x} + \dfrac{\partial \rho v^2 + p}{\partial y} = 0 \\[2mm] H_0 = \dfrac{\gamma p}{(\gamma - 1)\rho} + \dfrac{u^2 + v^2}{2} = \text{const} \end{cases} \qquad (9.1)$$

Note that the energy equation has been replaced by its first integral stating that the total enthalpy is constant. This holds for steady flow with uniform reservoir conditions, even when shocks are present. The equations are in conservation form.

Introducing a small perturbation hypothesis, for a symmetrical profile of vanishing thickness ϵ, the flow properties are expanded in series about a reference near-sonic state as

$$\begin{cases} \rho = \rho_0(1 + \epsilon^{\frac{2}{3}} \rho^{(1)} + \epsilon^{\frac{4}{3}} \rho^{(2)} + ...) \\[2mm] u = u_0(1 + \epsilon^{\frac{2}{3}} u^{(1)} + \epsilon^{\frac{4}{3}} u^{(2)} + ...) \\[2mm] v = u_0(\epsilon v^{(1)} + \epsilon^{\frac{5}{3}} v^{(2)} + ...) \\[2mm] p = p_0 + \rho_0 u_0^2(\epsilon^{\frac{2}{3}} p^{(1)} + \epsilon^{\frac{4}{3}} p^{(2)} + ...) \end{cases}$$

One further assumes that as $\epsilon \to 0$, the Mach number approaches unity, i.e., $1 - M_0^2 = O(\epsilon^{\frac{2}{3}})$, in order to obtain a mixed-type equation in the transformed plane (x, z), where $z = \epsilon^{\frac{1}{3}} y$.

The above expansions are carried into the Euler equations, and reordered by equal powers of ϵ. To the lowest order the following three equations are obtained:

$$\begin{cases} \dfrac{\partial v^{(1)}}{\partial x} - \dfrac{\partial u^{(1)}}{\partial z} = 0 \\[2mm] \rho^{(1)} = -u^{(1)} \\[2mm] p^{(1)} = -u^{(1)} \end{cases} \qquad (9.2)$$

In order to close the system, we need another equation for $u^{(1)}$ and $v^{(1)}$. It can be found at the next order where unknowns of second-order appear linearly and unknowns of first-order as nonlinear source terms. The unknowns of level two can be eliminated to give

$$\frac{\partial}{\partial x}\left(Ku^{(1)} - \frac{\gamma + 1}{2}\left(u^{(1)}\right)^2\right) + \frac{\partial v^{(1)}}{\partial z} = 0. \qquad (9.3)$$

K is the *transonic similarity parameter*. Its definition is not unique, up to a factor $M_0 \to 1$. Here are two commonly used definitions:

$$\begin{cases} K = \dfrac{1 - M_0^2}{\epsilon^{\frac{2}{3}}} \text{ (Cole)} \\[4mm] K = \dfrac{1 - M_0^2}{M_0^2 \epsilon^{\frac{2}{3}}} \text{ (Spreiter)} \end{cases}.$$

The first of equations (9.2) states that the flow is irrotational to first-order. There exists a perturbation potential $\varphi^{(1)}$ such that

$$\begin{cases} u^{(1)} = \dfrac{\partial \varphi^{(1)}}{\partial x} \\[4mm] v^{(1)} = \dfrac{\partial \varphi^{(1)}}{\partial z} \end{cases}.$$

Replacing the velocity components by the gradient of the potential in (9.3) yields the TSD equation in transformed variables:

$$\frac{\partial}{\partial x}\left(K \frac{\partial \varphi^{(1)}}{\partial x} - \frac{\gamma + 1}{2}\left(\frac{\partial \varphi^{(1)}}{\partial x}\right)^2 \right) + \frac{\partial^2 \varphi^{(1)}}{\partial z^2} = 0.$$

Note that this is the simplest nonlinear model equation that contains the fundamental features of transonic flows.

We can now rewrite the equation in the physical space prior to discretization. Dropping the upper index for simplicity, let $\varphi(x, y)$ be the potential of perturbation and $u(x, y)$, $v(x, y)$ the two components of $\nabla \varphi$. The TSD equation reads

$$\frac{\partial}{\partial x}\left((1 - M_0^2)\frac{\partial \varphi}{\partial x} - \frac{\gamma + 1}{2}M_0^2\left(\frac{\partial \varphi}{\partial x}\right)^2 \right) + \frac{\partial^2 \varphi}{\partial y^2} = 0. \qquad (9.4)$$

This second-order PDE can be transformed into a system of two first-order PDEs in terms of the perturbation velocity components, as

$$\begin{cases} \dfrac{\partial}{\partial x}\left((1 - M_0^2)\, u - \dfrac{\gamma + 1}{2}M_0^2 u^2 \right) + \dfrac{\partial v}{\partial y} = 0 \\[4mm] \dfrac{\partial v}{\partial x} - \dfrac{\partial u}{\partial y} = 0 \end{cases}. \qquad (9.5)$$

The boundary conditions associated with the TSD model for a symmetric profile in free air, are

$$\begin{cases} \dfrac{\partial \varphi}{\partial y}(x, 0) = v(x, 0) = f'(x) \\[4mm] \nabla \varphi = (u, v) \rightarrow (0, 0), \ x^2 + y^2 \rightarrow \infty \end{cases}.$$

The first condition is the *tangency condition* on the profile of equation $y = f(x)$. Note that the condition is applied on the x-axis, which is consistent with the small disturbance assumption.

In non-conservative form, equations (9.5) read

$$\begin{cases} \beta(u)\dfrac{\partial u}{\partial x} + \dfrac{\partial v}{\partial y} = 0 \\[2mm] \dfrac{\partial v}{\partial x} - \dfrac{\partial u}{\partial y} = 0 \end{cases},$$

where $\beta(u) = 1 - M_0^2 - (\gamma + 1)M_0^2 u$. Recast in matrix form

$$\begin{bmatrix} \beta(u) & 0 \\ 0 & 1 \end{bmatrix} \cdot \frac{\partial}{\partial x} \begin{bmatrix} u \\ v \end{bmatrix} + \begin{bmatrix} 0 & 1 \\ -1 & 0 \end{bmatrix} \cdot \frac{\partial}{\partial y} \begin{bmatrix} u \\ v \end{bmatrix} = 0,$$

the characteristic matrix identifies to

$$A = \begin{bmatrix} \beta(u)\dfrac{\partial \phi}{\partial x} & \dfrac{\partial \phi}{\partial y} \\[2mm] -\dfrac{\partial \phi}{\partial y} & \dfrac{\partial \phi}{\partial x} \end{bmatrix}.$$

The characteristic form follows:

$$Q = \beta(u) \left(\frac{\partial \phi}{\partial x} \right)^2 + \left(\frac{\partial \phi}{\partial y} \right)^2.$$

Let $u^* = ((1 - M_0^2)/(\gamma + 1)M_0^2)$ be the *sonic velocity* (or *critical velocity*). The local Mach number is $M = M_0 + (1 - M_0)(u/u^*)$.

Two situations are possible:

i) $u < u^* \Rightarrow \beta(u) > 0$, the equation is of elliptic type, the regime is *subsonic*

ii) $u > u^* \Rightarrow \beta(u) < 0$, the equation is of hyperbolic type, the flow regime is *supersonic*.

The TSD equation is of mixed elliptic-hyperbolic type. The transition from subsonic to supersonic can occur smoothly across a line $u = u^*$, the *sonic line*, or abruptly through a shock.

In the hyperbolic subdomain, two characteristic directions exist at each point. They correspond to the roots of $Q = 0$

$$\left(\frac{dy}{dx} \right)_{C^\pm} = \pm \frac{1}{\sqrt{-\beta(u)}}.$$

At the sonic line the characteristic lines have a cusp with vertical tangent. Let

$$\lambda = -\frac{2(-\beta(u))^{\frac{3}{2}}}{3(\gamma + 1)M_0^2} + v,$$

$$\mu = -\frac{2(-\beta(u))^{\frac{3}{2}}}{3(\gamma + 1)M_0^2} - v,$$

be the Riemann invarients along the characteristics C^+ and C^- respectively. The compatibility relations can be shown to be:

$$\begin{cases} \dfrac{\partial \lambda}{\partial x} + \dfrac{1}{\sqrt{-\beta(u)}} \dfrac{\partial \lambda}{\partial y} = 0 \\[2mm] \dfrac{\partial \mu}{\partial x} - \dfrac{1}{\sqrt{-\beta(u)}} \dfrac{\partial \mu}{\partial y} = 0. \end{cases}$$

Note that the Riemann invarient is constant along the corresponding characteristic.

The jump conditions are obtained from the conservation form (9.5), where n_x and n_y are the components of the unit normal vector to the jump line:

$$\begin{cases} \left\langle (1 - M_0^2)u - \frac{\gamma+1}{2} M_0^2 u^2 \right\rangle n_x + \langle v \rangle n_y = 0 \\[2mm] \langle v \rangle n_x - \langle u \rangle n_y = 0 \end{cases}.$$

Upon elimination of n_x and n_y one obtains the *shock polar*:

$$\beta(\bar{u}) \langle u \rangle^2 + \langle v \rangle^2 = 0,$$

where $\bar{u} = (u_1 + u_2)/2$.

The subscripts 1 and 2 represent the flow states before and after the shock. The shock polar for $M_0 = 1$, $\gamma = 1.4$ and $u_1 = 0.6$ is shown in Fig. 9.1. It is a semi-cubic.

Some remarkable points on the polar are, the point where the branches cross that corresponds to the supersonic state before the shock, $\langle u \rangle = \langle v \rangle = 0$, the point $\langle u \rangle = -4u_1/3$ that corresponds to the maximum deviation, $\langle v \rangle_{\max}$ possible with an oblique shock, and the point $\langle u \rangle = -2u_1$ that corresponds to the normal shock, $\langle v \rangle = 0$, the strongest possible shock at that Mach number.

The branches that extend beyond $u = u_1$, i.e. $\langle u \rangle > 0$, correspond to expansion shocks, which are not admissible.

The slope of the shock in the physical plane is given by

$$\left(\frac{dy}{dx} \right)_S = -\frac{n_x}{n_y} = -\frac{\langle u \rangle}{\langle v \rangle} = \pm \frac{1}{\sqrt{-\beta(\bar{u})}}.$$

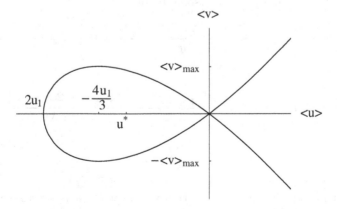

Fig. 9.1. Shock polar for the TSD equation

For shocks of vanishing strength, i.e. $\langle u \rangle \to 0$, the slope of the shock tends to the slope of the characteristic line.

9.3 The Murman–Cole Scheme (1970)

As was mentioned in the introduction, it was clear that in the subsonic domain SOR could be used efficiently and that in the supersonic domain a marching procedure would be required. However, the explicit scheme of Chap. 5 for the wave equation (5.7) could not be used because the stability requirement, $\Delta x \le |\beta(u)|\,\Delta y$, would not be satisfied near the sonic line. An essential ingredient was the choice of the implicit scheme (5.13) which allows us to march the solution on any mesh, thanks to its unconditional stability. This scheme required a simultaneous solution of the potential along a column in supersonic flow. This blended easily with the SLOR method at subsonic points (7.14).

The final ingredient is the test for switching scheme. In the first method, presented in 1970, Reference [5], Murman and Cole used three difference operators, a subsonic, a supersonic and a transition operator to allow subsonic flow to accelerate smoothly to supersonic speeds.

On a uniform mesh system, let $u_{i,j} = (\varphi_{i+1,j} - \varphi_{i-1,j})/2\Delta x$. The upper index is not indicated, but the latest available values of φ are used. The u's are called *switches*.

The scheme can be described by the following equations:

i) $u_{i-1,j} > u^*$ and $u_{i,j} > u^*$ (supersonic point):

$$\left((1 - M_0^2) - (\gamma + 1)M_0^2 \frac{\varphi_{i,j}^n - \varphi_{i-2,j}^{n+1}}{2\Delta x} \right) \frac{\varphi_{i,j}^{n+1} - 2\varphi_{i-1,j}^{n+1} + \varphi_{i-2,j}^{n+1}}{\Delta x^2}$$

$$+ \frac{\varphi_{i,j+1}^{n+1} - 2\varphi_{i,j}^{n+1} + \varphi_{i,j-1}^{n+1}}{\Delta y^2} = 0,$$

ii) $u_{i-1,j} < u^*$ and $u_{i,j} > u^*$ (sonic point):

$$\frac{\varphi_{i,j+1}^{n+1} - 2\varphi_{i,j}^{n+1} + \varphi_{i,j-1}^{n+1}}{\Delta y^2} = 0,$$

iii) $u_{i,j} < u^*$ (subsonic point):

$$\left((1 - M_0^2) - (\gamma + 1)M_0^2 \frac{\varphi_{i+1,j}^n - \varphi_{i-1,j}^{n+1}}{2\Delta x} \right) \frac{\varphi_{i+1,j}^n - 2\widetilde{\varphi}_{i,j} + \varphi_{i-1,j}^{n+1}}{\Delta x^2}$$

$$+ \frac{\widetilde{\varphi}_{i,j+1} - 2\widetilde{\varphi}_{i,j} + \widetilde{\varphi}_{i,j-1}}{\Delta y^2} = 0.$$

The final values are given by $\varphi^{n+1} = \varphi^n + \omega(\widetilde{\varphi} - \varphi^n)$.

Note that, at the sonic point, the nonlinear term is set to zero. With the inequalities holding, this algorithm enforces diagonal dominance for the tridiagonal matrix in all three cases. Subsonic points can be over-relaxed $(1 \leq \omega < 2)$.

This scheme is efficient, but the jump conditions are not satisfied at the foot of a shock on a smooth profile, where a normal shock is expected, that is one for which $u_2 = -u_1$. The error has been found by Murman to be attributable to a lack of conservation at the shock, due to the switching back from i) to iii). He corrected for this by having a new shock-point operator.

9.4 The Four-Operator Scheme of Murman (1973)

This four-operator scheme was designed to correct for the conservation error associated with the previous scheme. The details can be found in the paper of 1973 and Reference [6]. The new scheme reads:

i) $u_{i-1,j} > u^*$ and $u_{i,j} > u^*$ (supersonic point):

$$\left((1 - M_0^2) - (\gamma + 1)M_0^2 \frac{\varphi_{i,j}^n - \varphi_{i-2,j}^{n+1}}{2\Delta x} \right) \frac{\varphi_{i,j}^{n+1} - 2\varphi_{i-1,j}^{n+1} + \varphi_{i-2,j}^{n+1}}{\Delta x^2}$$

$$+ \frac{\varphi_{i,j+1}^{n+1} - 2\varphi_{i,j}^{n+1} + \varphi_{i,j-1}^{n+1}}{\Delta y^2} = 0,$$

ii) $u_{i-1,j} > u^*$ and $u_{i,j} < u^*$ (shock point):

$$\left((1 - M_0^2) - (\gamma + 1)M_0^2 \frac{\varphi_{i,j}^n - \varphi_{i-2,j}^{n+1}}{2\Delta x} \right) \frac{\varphi_{i,j}^{n+1} - 2\varphi_{i-1,j}^{n+1} + \varphi_{i-2,j}^{n+1}}{\Delta x^2}$$

$$+ \left((1 - M_0^2) - (\gamma + 1)M_0^2 \frac{\varphi_{i+1,j}^n - \varphi_{i-1,j}^{n+1}}{2\Delta x} \right) \frac{\varphi_{i+1,j}^n - 2\widetilde{\varphi}_{i,j} + \varphi_{i-1,j}^{n+1}}{\Delta x^2}$$

$$+ \frac{\widetilde{\varphi}_{i,j+1} - 2\widetilde{\varphi}_{i,j} + \widetilde{\varphi}_{i,j-1}}{\Delta y^2} = 0,$$

iii) $u_{i-1,j} < u^*$ and $u_{i,j} > u^*$ (sonic point):

$$\frac{\varphi_{i,j+1}^{n+1} - 2\varphi_{i,j}^{n+1} + \varphi_{i,j-1}^{n+1}}{\Delta y^2} = 0,$$

iv) $u_{i-1,j} < u^*$ and $u_{i,j} < u^*$ (subsonic point):

$$\left((1 - M_0^2) - (\gamma + 1)M_0^2 \frac{\varphi_{i+1,j}^n - \varphi_{i-1,j}^{n+1}}{2\Delta x} \right) \frac{\varphi_{i+1,j}^n - 2\widetilde{\varphi}_{i,j} + \varphi_{i-1,j}^{n+1}}{\Delta x^2}$$

$$+ \frac{\widetilde{\varphi}_{i,j+1} - 2\widetilde{\varphi}_{i,j} + \widetilde{\varphi}_{i,j-1}}{\Delta y^2} = 0.$$

Note that the shock point operator combines the x-derivatives of operators i) and iv) but the y-derivative is not doubled. A Taylor expansion of the TE indicates that this scheme is not consistent with the PDE, equation (9.4). However, the scheme is conservative and the jump conditions are satisfied in

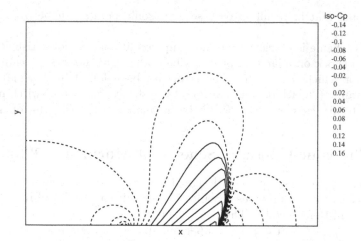

iso-Cp
-0.14
-0.12
-0.1
-0.08
-0.06
-0.04
-0.02
0
0.02
0.04
0.06
0.08
0.1
0.12
0.14
0.16

Fig. 9.2. Transonic flow field past a parabolic half-profile

the limit of fine meshes. The rationale here is that the notion of consistency is not relevant at a discontinuity, but conservation is of utmost importance. The sonic point is conservative, but not consistent, although the consistency error vanishes with the mesh step Δx, and it does not prevent expansion shocks to occur. A typical transonic flow field is plotted in Fig. 9.2. It represents the upper-half of the flow past a symmetric profile of equation $y = 0.0075x(1-x)$ placed in a uniform stream of air ($\gamma = 1.4$) with incoming Mach number $M = 0.94$. The reference Mach number is $M_0 = 1$. The iso-pressure lines, where $C_p = -2u$, are shown as dotted lines in the subsonic region and as solid lines for the supersonic bubble, including the sonic line $M = 1$. A curved shock is bounding the supersonic region on the downwind side.

Another example of transonic flow computed with the four-operator scheme of Murman is the symmetric transonic flow past the same 6% thick parabolic profile at $M = 1.17$. The iso-pressure lines are shown in Fig. 9.3. The shock is detached from the leading edge and a small subsonic bubble is embedded in an overall supersonic flow. The subsonic iso-pressure lines are shown as broken lines, while the supersonic iso-pressure lines are shown as solid lines. Note that, as the shock wave weakens away from the profile, it tends to a characteristic line.

To give some insight in the way the shock point is enforcing the correct solution, we will consider the one-dimensional shock problem. The y-derivative vanishes and, with $M_0 = 1$ the equation reduces to

$$\frac{d}{dx}\left(\frac{1}{2}\left(\frac{d\varphi}{dx}\right)^2\right) = 0, \ 0 \leq x \leq 1. \tag{9.6}$$

This is Burgers' equation (inviscid) for $u = d\varphi/dx$. Note also that $u^* = 0$.

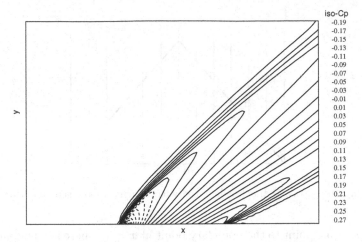

Fig. 9.3. Transonic flow past a parabolic half-profile

The boundary conditions are:

$$\begin{cases} \varphi(0) = 0, \ \dfrac{d\varphi}{dx}(0) = u_0 > 0 \\ \varphi(1) = 0 \end{cases}.$$ (9.7)

Note that for this second-order ODE, three conditions are needed, the first two prescribe the supersonic incoming flow, the third fixes the position of the shock at $x = \frac{1}{2}$. The jump condition is

$$\overline{\dfrac{d\varphi}{dx}} = \frac{1}{2}\left[\left(\frac{d\varphi}{dx}\right)_1 + \left(\frac{d\varphi}{dx}\right)_2\right] = 0.$$ (9.8)

At convergence the solution must satisfy the following FDEs:
i) $u_{i-1} > 0$ and $u_i > 0$ (supersonic point):

$$\left(\frac{\varphi_i - \varphi_{i-2}}{2\Delta x}\right)\frac{\varphi_i - 2\varphi_{i-1} + \varphi_{i-2}}{\Delta x^2} = 0,$$

ii) $u_{i-1} < 0$ and $u_i > 0$ (sonic point):

$$\left(\frac{\varphi_{i+1} - \varphi_{i-1}}{2\Delta x}\right)\frac{\varphi_i - 2\varphi_{i-1} + \varphi_{i-2}}{\Delta x^2} = 0,$$

iii) $u_i < 0$ (subsonic point):

$$\left(\frac{\varphi_{i+1} - \varphi_{i-1}}{2\Delta x}\right)\frac{\varphi_{i+1} - 2\varphi_i + \varphi_{i-1}}{\Delta x^2} = 0.$$

The first equation indicates that all the supersonic points are aligned on the exact supersonic branch. There is no sonic point in this problem. Similarly, all the subsonic points are aligned on a straight line connecting the

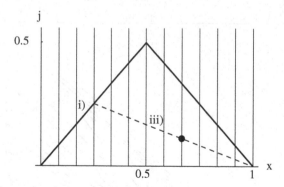

Fig. 9.4. Normal shock calculated with Murman-Cole scheme (1970)

last supersonic point to the boundary point at $x = 1$. There is no requirement in the scheme concerning the slope of the subsonic branch, so long as it is negative. Thus, the subsonic branch slope can vary between $-u_0$ and zero, in other words the shock location can be at any node i_s, $2 \leq i_s \leq (ix+1)/2$. In Fig. 9.4 the exact solution is shown as the solid line and a solution admitted by the Murman-Cole scheme as a broken line. With this scheme the shock is always located at a node.

With the conservative scheme, however, the following FDEs must be satisfied:

i) $u_{i-1} > 0$ and $u_i > 0$(supersonic point):

$$\left(\frac{\varphi_i - \varphi_{i-2}}{2\Delta x}\right)\frac{\varphi_i - 2\varphi_{i-1} + \varphi_{i-2}}{\Delta x^2} = 0,$$

ii) $u_{i-1} > 0$ and $u_i < 0$ (shock point):

$$\left(\frac{\varphi_i - \varphi_{i-2}}{2\Delta x}\right)\frac{\varphi_i - 2\varphi_{i-1} + \varphi_{i-2}}{\Delta x^2} + \left(\frac{\varphi_{i+1} - \varphi_{i-1}}{2\Delta x}\right)\frac{\varphi_{i+1} - 2\varphi_i + \varphi_{i-1}}{\Delta x^2} = 0,$$

iii) $u_{i-1} < 0$ and $u_i > 0$ (sonic point):

$$\left(\frac{\varphi_{i+1} - \varphi_{i-1}}{2\Delta x}\right)\frac{\varphi_i - 2\varphi_{i-1} + \varphi_{i-2}}{\Delta x^2} = 0,$$

iv) $u_{i-1} < 0$ and $u_i < 0$ (subsonic point):

$$\left(\frac{\varphi_{i+1} - \varphi_{i-1}}{2\Delta x}\right)\frac{\varphi_{i+1} - 2\varphi_i + \varphi_{i-1}}{\Delta x^2} = 0.$$

The supersonic points are still on the exact supersonic branch. The sonic point operator is not used. The subsonic points are all aligned. The key element is the shock point operator ii). It can be written in conservation form as:

$$\frac{\frac{1}{2}\left(\frac{\varphi_i - \varphi_{i-1}}{\Delta x}\right)^2 - \frac{1}{2}\left(\frac{\varphi_{i-1} - \varphi_{i-2}}{\Delta x}\right)^2}{\Delta x}$$

$$+ \frac{\frac{1}{2}\left(\frac{\varphi_{i+1} - \varphi_i}{\Delta x}\right)^2 - \frac{1}{2}\left(\frac{\varphi_i - \varphi_{i-1}}{\Delta x}\right)^2}{\Delta x} = 0.$$

This is actually the original approach from which the schemes for the nonlinear term are derived. After cancellation of the first and last terms and factorization one gets

$$\frac{\frac{1}{2}\left(\frac{\varphi_{i+1} - \varphi_i}{\Delta x} + \frac{\varphi_{i-1} - \varphi_{i-2}}{\Delta x}\right)\left(\frac{\varphi_{i+1} - \varphi_i}{\Delta x} - \frac{\varphi_{i-1} - \varphi_{i-2}}{\Delta x}\right)}{\Delta x} = 0.$$

Note that the second bracket can be expressed in terms of the switches

$$\frac{\frac{1}{2}\left(\frac{\varphi_{i+1} - \varphi_i}{\Delta x} + \frac{\varphi_{i-1} - \varphi_{i-2}}{\Delta x}\right)(u_i - u_{i-1})}{\Delta x} = 0.$$

Since, at the shock point, $u_{i-1} > 0$ and $u_i < 0$, the term in the last parenthesis cannot be zero. The converged solution therefore must satisfy at the shock point

$$\frac{1}{2}\left(\frac{\varphi_{i-1} - \varphi_{i-2}}{\Delta x} + \frac{\varphi_{i+1} - \varphi_i}{\Delta x}\right) = 0,$$

which is condition (9.8). This is shown in Fig. 9.5.

Note, that, unlike the previous scheme, the shock location can be traced to its exact position, here $x = \frac{1}{2}$, independently of the mesh system used.

The convergence of the four operator scheme is displayed in Fig. 9.6. The mesh consists of only nine points, but the process is independent of the mesh

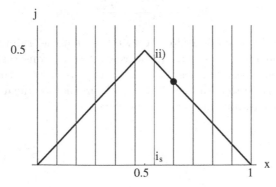

Fig. 9.5. Normal shock calculated with Murman's conservative scheme (1973)

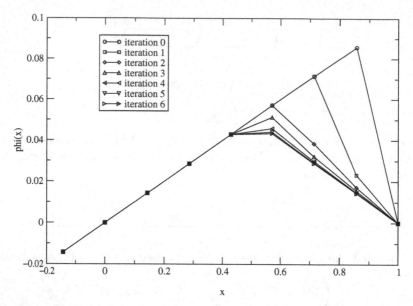

Fig. 9.6. Normal shock calculated with Murman four-operator scheme

size, except for the number of iterations required to reach steady-state. Here it takes six iterations to converge. Note that the first mesh point is located at $x_1 = -\Delta x$ to enforce the initial condition. The first point calculated is $i = 3$.

Problem: Show that the following scheme can replace the sonic point scheme in the four-operator scheme of Murman

iii) $u_{i-1,j} < u^*$ and $u_{i,j} > u^*$ (sonic point):

$$\left((1 - M_0^2) - (\gamma + 1)M_0^2 \frac{\varphi_{i,j}^{n+1} - \varphi_{i-1,j}^{n+1}}{\Delta x} \right) \frac{\varphi_{i+1,j}^n - \varphi_{i,j}^n - \varphi_{i-1,j}^{n+1} + \varphi_{i-2,j}^{n+1}}{2\Delta x^2}$$

$$+ \frac{\varphi_{i,j+1}^{n+1} - 2\varphi_{i,j}^{n+1} + \varphi_{i,j-1}^{n+1}}{\Delta y^2} = 0.$$

Show, using the inequalities for the switches, that the tridiagonal matrix is diagonally dominant. This scheme is consistent but not conservative, although the conservation error vanishes with Δx, and it prevents expansion shocks from occuring. Note that the new value $\varphi_{i,j}^{n+1}$ in the nonlinear term is taken from the coefficient, not the second derivative.

10. Treatment of Non-Linearities

10.1 Introduction

Three nonlinear model problems have been discussed in previous chapters, the Burgers' equation (inviscid and viscous) and the transonic small disturbance equation (TSD). Explicit as well as implicit schemes have been introduced to solve these equations, using time-marching and relaxation algorithms. In all cases, a linear equation or system for the new values was solved to update the solution.

In fact, the nonlinear algebraic system obtained by discretization of a nonlinear PDE is always linearized in the variables u^{n+1} or φ^{n+1} for the purpose of advancing the state, and the nonlinear solution is obtained in the limit when a steady-state is reached.

There are several techniques of linearization, and the linearized sets obtained are not unique.

We will investigate the mixed-type scheme for Burgers' equation (inviscid) to understand the role played by the nonlinear term, first using an explicit scheme, then an implicit scheme.

10.2 An Explicit Mixed-Type Scheme

Let $u(x,t)$ be the solution of Burgers' equation with initial and boundary conditions

$$\begin{cases} \dfrac{\partial u}{\partial t} + \dfrac{\partial}{\partial x}\left(\dfrac{u^2}{2}\right) = 0,\ 0 \le x \le 1,\ t \ge 0 \\[2mm] u(x,0) = 1 - 2x,\ 0 \le x \le 1 \\[2mm] u(0,t) = 1,\ u(1,t) = -1,\ t \ge 0 \end{cases} \qquad (10.1)$$

This corresponds to the shock formation problem. The explicit scheme reads

i) $u_{i-\frac{1}{2}}^n > 0,\ u_{i+\frac{1}{2}}^n > 0$ (supersonic point):

$$\frac{u_i^{n+1} - u_i^n}{\Delta t} + u_{i-\frac{1}{2}}^n \frac{u_i^n - u_{i-1}^n}{\Delta x} = 0,$$

ii) $u_{i-\frac{1}{2}}^n > 0$, $u_{i+\frac{1}{2}}^n < 0$ (shock point):

$$\frac{u_i^{n+1} - u_i^n}{\Delta t} + u_{i-\frac{1}{2}}^n \frac{u_i^n - u_{i-1}^n}{\Delta x} + u_{i+\frac{1}{2}}^n \frac{u_{i+1}^n - u_i^n}{\Delta x} = 0,$$

iii) $u_{i-\frac{1}{2}}^n < 0$, $u_{i+\frac{1}{2}}^n > 0$ (sonic point):

$$\frac{u_i^{n+1} - u_i^n}{\Delta t} + u_i^{n+1} \frac{u_{i+1}^n - u_{i-1}^n}{2\Delta x} = 0,$$

iv) $u_{i-\frac{1}{2}}^n < 0$, $u_{i+\frac{1}{2}}^n < 0$ (subsonic point):

$$\frac{u_i^{n+1} - u_i^n}{\Delta t} + u_{i+\frac{1}{2}}^n \frac{u_{i+1}^n - u_i^n}{\Delta x} = 0,$$

where the mid-value correspond to averages as

$$u_{i-\frac{1}{2}} = \frac{u_{i-1} + u_i}{2}.$$

The space discretization in schemes i), ii) and iv) corresponds to the Murman (1973) scheme when one lets $u_i = (\varphi_i - \varphi_{i-1})/\Delta x$. Scheme iii) represents the sonic acceleration and by construction allows for a smooth subsonic-supersonic transition (see Reference [7]).

The study of consistency and accuracy is carried out in the same way for linear and nonlinear PDEs. Let $F(u) = u^2/2$ be the flux function. Schemes i), ii) and iv) can be formulated in conservation FDE form as, e.g. ii)

$$\frac{u_i^{n+1} - u_i^n}{\Delta t} + \frac{F_i^n - F_{i-1}^n}{\Delta x} + \frac{F_{i+1}^n - F_i^n}{\Delta x} = 0. \tag{10.2}$$

The TE reads:

$$\epsilon_i^n = \frac{\partial u_i^n}{\partial t} + \frac{\Delta t}{2} \frac{\partial^2 u_i^n}{\partial t^2} + O(\Delta t^2) + \frac{\partial F_i^n}{\partial x} - \frac{\Delta x}{2} \frac{\partial^2 F_i^n}{\partial x^2} + \frac{\partial F_i^n}{\partial x} + \frac{\Delta x}{2} \frac{\partial^2 F_i^n}{\partial x^2} + O(\Delta x^2)$$

$$= \left(\frac{\partial u_i^n}{\partial t} + 2 \frac{\partial F_i^n}{\partial x} \right) + O(\Delta t, \Delta x).$$

The shock-point operator is not consistent with the PDE, as mentioned earlier. It enforces the conservation property of the scheme, which is more important.

The above result indicates that schemes i) and iv) are consistent and first-order accurate in t and x.

Scheme iii) is clearly first-order in t and second-order in x, since

$$\epsilon_i^n = \Delta t \left(\frac{1}{2} \frac{\partial^2 u_i^n}{\partial t^2} + \frac{\partial u_i^n}{\partial t} \frac{\partial u_i^n}{\partial x} \right) + \frac{\Delta x^2}{3!} u_i^n \frac{\partial^3 u_i^n}{\partial x^3} + O(\Delta t^2, \Delta t \Delta x^2, \Delta x^4)$$

$$= O(\Delta t, \Delta x^2).$$

It is not conservative, however, one can show that the conservation error goes to zero with vanishing mesh steps.

The study of stability is of great interest since we deal with a nonlinear problem. The FDE must be linearized before the Von Neumann analysis can be used. Consider the shock-point operator (10.2) which encompasses schemes i) and iv). If the term F_i^n is removed, the linearization will produce

$$\frac{u_i^{n+1} - u_i^n}{\Delta t} + \left(u_{i-1}^n + u_{i+1}^n \right) \frac{u_{i+1}^n - u_{i-1}^n}{2\Delta x} = 0.$$

Freezing the coefficient by letting $c = \left(u_{i-1}^n + u_{i+1}^n \right)$ leads to a centered scheme for the linear convection equation (5.1) that is unconditionally unstable. Since this scheme is known to produce good results, we have to revisit the linearization. The important feature of the shock point is the change of sign of the characteristic speed $u_{i\pm\frac{1}{2}}^n$. A linearization that accounts for this feature is

$$\frac{u_i^{n+1} - u_i^n}{\Delta t} + c^+ \frac{u_i^n - u_{i-1}^n}{\Delta x} + c^- \frac{u_{i+1}^n - u_i^n}{\Delta x} = 0,$$

where $c^+ = u_{i-\frac{1}{2}}^n > 0$, and $c^- = u_{i+\frac{1}{2}}^n < 0$. Now, proceeding with Von Neumann analysis, with Courant numbers $\sigma^+ = c^+(\Delta t/\Delta x) > 0$, $\sigma^- = -c^-(\Delta t/\Delta x) > 0$, and $u_i^n = g^n e^{ii\alpha}$, we obtain

$$g = 1 - \sigma^+ \left(1 - e^{-i\alpha} \right) + \sigma^- \left(e^{i\alpha} - 1 \right)$$

$$= 1 - \left(\sigma^- + \sigma^+ \right) \left(1 - \cos\alpha \right) + i \left(\sigma^- - \sigma^+ \right) \sin\alpha.$$

The condition $|g|^2 \leq 1$ is equivalent to $(\sigma^- + \sigma^+)(1 - (\sigma^- + \sigma^+)) \geq 0$. The stability condition for the shock-point operator is:

$$\sigma^- + \sigma^+ \leq 1.$$

In terms of the time step, this is
ii)

$$\Delta t \leq \frac{\Delta x}{u_{i-\frac{1}{2}}^n - u_{i+\frac{1}{2}}^n}.$$

The nonlinear feature explains why this "centered" scheme is stable.

Problem: Verify the result of the stability analysis by computing $|g|^2$ and discussing its magnitude.

The stability conditions for the supersonic- and subsonic-point operators are obtained immediately as

i)

$$\Delta t \le \frac{\Delta x}{u^n_{i-\frac{1}{2}}},$$

iv)

$$\Delta t \le -\frac{\Delta x}{u^n_{i+\frac{1}{2}}}.$$

Finally, the unconditional stability of the sonic point is easily explained by its "implicitness". Let $\Delta = \Delta t((u^n_{i+1} - u^n_{i-1})/2\Delta x > 0)$. The amplification factor is found to be

$$g = \frac{1}{1+\Delta} \le 1.$$

Note that the linearization "freezes" the partial derivative, not the coefficient, as is usually the case. Here the feature that must be passed on to the Von Neumann analysis is the fact that the coefficient of the partial derivative $\partial u/\partial x$ is the new value u^{n+1}_i, making the scheme "locally implicit".

10.3 An Implicit Mixed-Type Scheme

The previous scheme can be made implicit by shifting the space derivatives in i), ii) and iv) to level $n+1$:

i) $u^n_{i-\frac{1}{2}} > 0$, $u^n_{i+\frac{1}{2}} > 0$ (supersonic point):

$$\frac{u^{n+1}_i - u^n_i}{\Delta t} + u^n_{i-\frac{1}{2}} \frac{u^{n+1}_i - u^{n+1}_{i-1}}{\Delta x} = 0,$$

ii) $u^n_{i-\frac{1}{2}} > 0$, $u^n_{i+\frac{1}{2}} < 0$ (shock point):

$$\frac{u^{n+1}_i - u^n_i}{\Delta t} + u^n_{i-\frac{1}{2}} \frac{u^{n+1}_i - u^{n+1}_{i-1}}{\Delta x} + u^n_{i+\frac{1}{2}} \frac{u^{n+1}_{i+1} - u^{n+1}_i}{\Delta x} = 0,$$

iii) $u^n_{i-\frac{1}{2}} < 0$, $u^n_{i+\frac{1}{2}} > 0$ (sonic point):

$$\frac{u^{n+1}_i - u^n_i}{\Delta t} + u^{n+1}_i \frac{u^n_{i+1} - u^n_{i-1}}{2\Delta x} = 0,$$

iv) $u^n_{i-\frac{1}{2}} < 0$, $u^n_{i+\frac{1}{2}} < 0$ (subsonic point):

$$\frac{u^{n+1}_i - u^n_i}{\Delta t} + u^n_{i+\frac{1}{2}} \frac{u^{n+1}_{i+1} - u^{n+1}_i}{\Delta x} = 0.$$

Note that the sonic-point does not need to be changed. This scheme is unconditionally stable and will converge to steady-state for $0 < \Delta t < \infty$. When Δt is very large (say $\Delta t = 10^6$), the scheme becomes an SLOR scheme for the steady Burgers' equation.

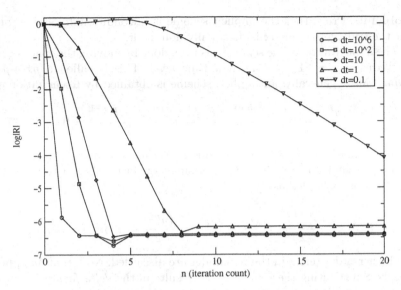

Fig. 10.1. Convergence to steady-state of the implicit algorithm for a range of time steps

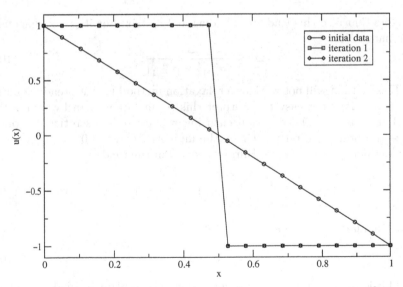

Fig. 10.2. Evolution of the solution with iterations for $\Delta t = 10^6$

Convergence curves are presented in Fig. 10.1 for a range of time steps $\Delta t = 10^{-1}$, 1, 10, 10^2, 10^6. For time steps above 10^6 the method converges in essentially two iterations as shown in Fig. 10.2.

Problem: Prove that the implicit scheme is unconditionally stable. Show that the tridiagonal matrix is diagonally dominant.

The implicitation of the scheme has been done by moving the space derivative terms in the FDE to the new time level. This is called a *fixed-point method*. More generally, an implicit scheme is obtained by linearization of

$$\frac{u_i^{n+1} - u_i^n}{\Delta t} + \frac{F_i^{n+1} - F_{i-1}^{n+1}}{\Delta x} + \frac{F_{i+1}^{n+1} - F_i^{n+1}}{\Delta x} = 0.$$

A systematic linearization technique is *Newton's method*.

Assume that $\delta u = u^{n+1} - u^n$ is a small quantity, then a term such as F_i^{n+1} can be expanded as

$$F_i^{n+1} = \frac{\left(u_i^{n+1}\right)^2}{2} = \frac{\left(u_i^n + \delta u_i\right)^2}{2} = \frac{\left(u_i^n\right)^2}{2} + u_i^n \delta u_i + O(\delta u_i)^2.$$

The higher-order terms in the remainder are discarded. Note that δu appears linearly. Substituting these in the FDE results in the *delta-form*:

$$-\frac{u_{i-1}^n}{\Delta x}\delta u_{i-1} + \frac{1}{\Delta t}\delta u_i + \frac{u_{i+1}^n}{\Delta x}\delta u_{i+1} = -\frac{F_i^n - F_{i-1}^n}{\Delta x} - \frac{F_{i+1}^n - F_i^n}{\Delta x}.$$

Note that also this leads to a tridiagonal matrix, but it is not diagonally dominant unless

$$\Delta t \leq \frac{\Delta x}{\left|u_{i-1}^n\right| + \left|u_{i+1}^n\right|}. \tag{10.3}$$

This method will not work as a relaxation method for the steady equation ($\Delta t = \infty$). It is not easy to interpret this result, other than by saying that (10.3) acts as a CFL condition ensuring a consistent discretization of the time-evolution term in the PDE in the limit as $\Delta t, \Delta x \to 0$.

Consider now the steady Burger's equation (inviscid)

$$\frac{d}{dx}\left(\frac{u^2}{2}\right) = 0, \; 0 \leq x \leq 1, \tag{10.4}$$

with boundary conditions

$$u(0) = 1, \; u(1) = -1. \tag{10.5}$$

This is the counterpart of the normal shock problem studied in Chap. 9, but the difference with the potential formulation (9.6)–(9.7) is that the solution to (10.4)–(10.5) is not unique. The shock can be placed at any position in $]0, 1[$. However, some iterative procedure, viewed as a time-evolution process, with an initial condition, may make the shock position unique, in the same way that the unsteady Burgers' equation with boundary conditions and initial condition, equations (10.1), will converge to a unique shock location.

10.4 Discussion

Before we conclude this chapter, it is important to stress that not many theorems have been proven for nonlinear PDEs, as far as the properties of existence, uniqueness and being well posed of the problem are concerned. Nonetheless, we have seen that the TE and the Von Neumann analysis remain extremely useful tools to assess fundamental properties of a scheme. Often, some extra physical insight will be of great help in designing a numerical scheme and selecting a solution algorithm.

When it comes to choosing a linearization technique, experience may be the best guide. Nonlinear problems are likely to have multiple solutions, some of which may not be physically relevant. As a steady iterative method, Newton's method fails because it recognizes the fact that the solution is not unique. When Newton's method can be used, however, quadratic convergence is obtained. To illustrate this, consider the family of iterative procedures, or sequences, to find the positive root to the equation $u^2 = a > 0$:

$$\begin{cases} u^{n+1} = u^n + \frac{\omega}{2}\left(\frac{a}{u^n} - u^n\right), \ u^0 > 0 \\ 0 < \omega \le 1 \end{cases}.$$

$\omega = 1$ corresponds to Newton's method. To grasp the difference that occurs when $\omega = 1$, let's assume that the sequence converges and introduce the normalized error

$$\epsilon^n = \frac{u^n - \sqrt{a}}{\sqrt{a}}.$$

The recurrence relation for ϵ^{n+1} is

$$\epsilon^{n+1} = \frac{2\left(1 - \omega\right)\epsilon^n + \left(2 - \omega\right)\left(\epsilon^n\right)^2}{2\left(1 + \epsilon^n\right)}.$$

As $n \to \infty$, $\epsilon^n \to 0$, two cases are possible:

i) $\omega < 1$, then $\epsilon^{n+1} \approx (1 - \omega)\epsilon^n$, the amplification factor is $g = 1 - \omega < 1$, the convergence is *geometric*,

ii) $\omega = 1$, then $\epsilon^{n+1} \approx \frac{(\epsilon^n)^2}{2}$, the amplification factor is $g = \frac{\epsilon^n}{2} \to 0$, the convergence is *quadratic*.

Both cases are shown in Fig. 10.3.

Problem: Consider the following nonlinear second-order ODE:

$$\begin{cases} \left(1 + u^2\right)u'' = -2\left(1 + x^2\left(1 - x\right)^2\right), \ 0 \le x \le 1 \\ u(0) = u(1) = 0 \end{cases}.$$

Use an implicit scheme to solve the set of algebraic equations:

i) using a fixed-point linearization (freeze the coefficient $\left(1 + (u_i^n)^2\right)$ to the old value),

ii) using Newton's method of linearization.

Check your result against those shown in the plots.

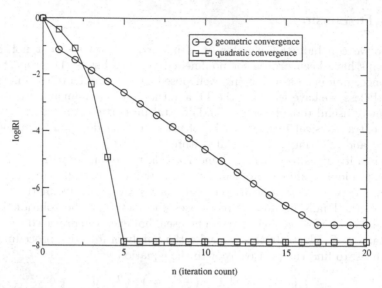

Fig. 10.3. Geometric ($\omega = 0.5$) and quadratic ($\omega = 1$) convergence of a sequence.

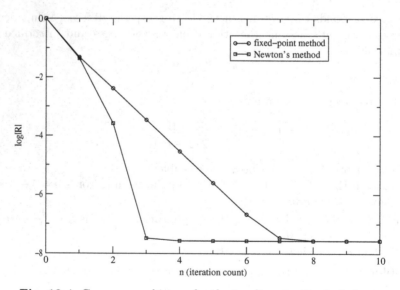

Fig. 10.4. Convergence history for the two linearization techniques

In Fig. 10.4 the convergence of both methods are shown for a mesh of $ix = 21$ points and an initial condition $u_0(x) = 0$. The fixed-point method yields a geometric convergence, whereas Newton's method yields quadratic convergence.

11. Application to a System of Equations

11.1 Introduction

The quasi-one-dimensional Euler equations, or gas dynamics equations, are of great practical use to represent phenomena taking place in slowly varying channels and ducts. For large Reynolds numbers, the viscous effects can be neglected, and the result will be useful for understanding steady flow in a converging-diverging nozzle, or unsteady flow in a shock tube. As a time-dependent system, we will see that it is hyperbolic and that the solution can be marched in time. As a steady system, it is a challenging coupled set of nonlinear ODEs of first-order, with singular points. For that reason, it will always be preferable to use the unsteady approach and solve for the steady flows in the limit of very large times.

11.2 The Equations of Gas Dynamics

The equations governing the conservation of mass, axial momentum and energy, in a slowly varying nozzle of cross section $g(x)$ are

$$
\begin{cases}
\dfrac{\partial \rho g}{\partial t} + \dfrac{\partial \rho u g}{\partial x} = r_1 = 0 \\[2ex]
\dfrac{\partial \rho u g}{\partial t} + \dfrac{\partial \left(\rho u^2 + p\right) g}{\partial x} - p g' = r_2 = 0 \\[2ex]
\dfrac{\partial \rho E g}{\partial t} + \dfrac{\partial \rho u H g}{\partial x} = r_3 = 0
\end{cases}
\qquad (11.1)
$$

$E = e + (u^2/2) = (p/(\gamma - 1)\rho) + (u^2/2)$ is the total energy, $H = h + (u^2/2) = (\gamma p/(\gamma - 1)\rho) + (u^2/2)$ is the total enthalpy. This system is in conservation form. Using a more compact vector notation, this is

$$
\frac{\partial w}{\partial t} + \frac{\partial F(w)}{\partial x} + G(w) = 0,
$$

where

$$w = \begin{bmatrix} \rho g \\ \rho u g \\ \rho E g \end{bmatrix} \quad F(w) = \begin{bmatrix} \rho u g \\ (\rho u^2 + p)g \\ \rho u H g \end{bmatrix} \quad G(w) = \begin{bmatrix} 0 \\ -pg' \\ 0 \end{bmatrix}.$$

The type of this system is studied in the quasi-linear form of the system, a form in which the derivatives appear linearly, expressed in terms of three independent unknowns. Here we choose the three components of the vector w, i.e. w_1, w_2 and w_3. Note that the nozzle area $g(x)$ does not influence the type of the system, since in the quasi-linear form it does not appear as a coefficient of the partial derivatives, but as a source term. Without restriction we will let $g = \text{const} = 1$. Eliminating pressure from the system yields

$$F(w) = \begin{bmatrix} w_2 \\ \dfrac{3-\gamma}{2}\dfrac{w_2^2}{w_1} + (\gamma-1)w_3 \\ -\dfrac{\gamma-1}{2}\dfrac{w_2^3}{w_1^2} + \gamma\dfrac{w_2 w_3}{w_1} \end{bmatrix}.$$

Using the chain rule, the quasi-linear form can be expressed as

$$\frac{\partial w}{\partial t} + \frac{\partial F}{\partial w}\cdot\frac{\partial w}{\partial x} = 0,$$

where $\partial F/\partial w$ is the *Jacobian matrix*. Expressed in terms of the w_i's it reads

$$\frac{\partial F}{\partial w} = \begin{bmatrix} 0 & 1 & 0 \\ -\dfrac{3-\gamma}{2}\dfrac{w_2^2}{w_1^2} & (3-\gamma)\dfrac{w_2}{w_1} & \gamma-1 \\ (\gamma-1)\dfrac{w_2^3}{w_1^3} - \gamma\dfrac{w_2 w_3}{w_1^2} & -3\dfrac{\gamma-1}{2}\dfrac{w_2^2}{w_1^2} + \gamma\dfrac{w_3}{w_1} & \gamma\dfrac{w_2}{w_1} \end{bmatrix},$$

or introducing the speed of sound $a^2 = \frac{\gamma p}{\rho}$, the Jacobian matrix simplifies to

$$\frac{\partial F}{\partial w} = \begin{bmatrix} 0 & 1 & 0 \\ -\dfrac{3-\gamma}{2}u^2 & (3-\gamma)u & \gamma-1 \\ -\dfrac{2-\gamma}{2}u^3 - \dfrac{a^2 u}{\gamma-1} & \dfrac{3-2\gamma}{2}u^2 + \dfrac{a^2}{\gamma-1} & \gamma u \end{bmatrix}.$$

The characteristic matrix is now obtained:

$$A = \begin{bmatrix} \dfrac{\partial\phi}{\partial t} & \dfrac{\partial\phi}{\partial x} & 0 \\ -\dfrac{3-\gamma}{2}u^2\dfrac{\partial\phi}{\partial x} & \dfrac{\partial\phi}{\partial t} + (3-\gamma)u\dfrac{\partial\phi}{\partial x} & (\gamma-1)\dfrac{\partial\phi}{\partial x} \\ -\left(\dfrac{2-\gamma}{2}u^3 + \dfrac{a^2 u}{\gamma-1}\right)\dfrac{\partial\phi}{\partial x} & \left(\dfrac{3-2\gamma}{2}u^2 + \dfrac{a^2}{\gamma-1}\right)\dfrac{\partial\phi}{\partial x} & \dfrac{\partial\phi}{\partial t} + \gamma u\dfrac{\partial\phi}{\partial x} \end{bmatrix},$$

and the characteristic form, after some algebra, can be written as

$$Q = \left(\frac{\partial \phi}{\partial t} + u \frac{\partial \phi}{\partial x}\right) \left(\left(\frac{\partial \phi}{\partial t} + u \frac{\partial \phi}{\partial x}\right)^2 - a^2 \left(\frac{\partial \phi}{\partial x}\right)^2\right).$$

There are three distinct characteristic wave speeds, which are the eigenvalues of the Jacobian matrix and the roots of the characteristic form

$$\left(\frac{dx}{dt}\right)_C = -\frac{\frac{\partial \phi}{\partial t}}{\frac{\partial \phi}{\partial x}} = \lambda_j = \begin{cases} u - a, \ C^- \\ u, \ C^0 \\ u + a, \ C^+ \end{cases}, \ j = 1, 2, 3.$$

The system is totally hyperbolic. The two characteristics $C^{\pm}$ correspond to acoustic waves, C^0 corresponds to the particle path. The compatibility relations are obtained by replacing one column of A by the terms (r_1, r_2, r_3) and setting the determinant to zero. One gets:

$$\begin{cases} CR^- : \left(\frac{\gamma-1}{2}u^2 + au\right) r_1 - \left(a + (\gamma-1)u\right) r_2 + (\gamma-1) r_3 = 0 \\ CR^0 : \left(\frac{\gamma-1}{2}u^2 - a^2\right) r_1 - (\gamma-1) u r_2 + (\gamma-1) r_3 = 0 \\ CR^+ : \left(\frac{\gamma-1}{2}u^2 - au\right) r_1 + \left(a - (\gamma-1)u\right) r_2 + (\gamma-1) r_3 = 0 \end{cases} \quad . \quad (11.2)$$

The coefficients of (r_1, r_2, r_3) in each compatibility relation are the components of the *left-eigenvectors* of the Jacobian matrix, i.e.

$$(l_1^{(j)}, l_2^{(j)}, l_3^{(j)}).(r_1, r_2, r_3)^t = 0.$$

They satisfy the identity

$$l^{(j)}.\frac{\partial F}{\partial w} = l^{(j)} \lambda_j.$$

Problem: The flux vector terms $f(w)$ in the Euler equations are homogeneous functions of degree one of the w_i's. A homogeneous function of degree n is such that $f(kw) = k^n f(w)$. Show that, for such functions, the following property holds:

$$f(w) = \frac{\partial f}{\partial w}.w.$$

11.3 Jump Conditions

The conservation form (11.1) is used to derive the jump conditions

$$\begin{cases} \langle \rho g \rangle n_t + \langle \rho u g \rangle n_x = 0 \\ \langle \rho u g \rangle n_t + \langle (\rho u^2 + p) g \rangle n_x = 0 \\ \langle \rho E g \rangle n_t + \langle \rho u H g \rangle n_x = 0 \end{cases} .$$

In most problems the area $g(x)$ is continuous, therefore $\langle g \rangle = 0$ and the jump conditions reduce to

$$
\begin{cases}
\langle \rho \rangle \, n_t + \langle \rho u \rangle \, n_x = 0 \\
\langle \rho u \rangle \, n_t + \langle \rho u^2 + p \rangle \, n_x = 0 \ . \\
\langle \rho E \rangle \, n_t + \langle \rho u H \rangle \, n_x = 0
\end{cases}
$$

There are two types of jump lines, those crossed by the fluid and those along a particle path.

If the fluid crosses a jump line, i.e. $u_1, u_2 > -(n_t/n_x)$ or $u_1, u_2 < -(n_t/n_x)$, there is a discontinuity in the velocity, density and pressure, $\langle u \rangle, \langle \rho \rangle, \langle p \rangle \neq 0$. The discontinuity is called a *shock*.

If the fluid does not cross the discontinuity, i.e. $u_1 = u_2 = -(n_t/n_x)$, the pressure is continuous across the jump line, $\langle p \rangle = 0$, the jump in density cannot be determined from the jump conditions. The discontinuity is called a *contact discontinuity*.

In steady flow, the jump conditions reduce even further to

$$
\begin{cases}
\langle \rho u \rangle = 0 \\
\langle \rho u^2 + p \rangle = 0 \ . \\
\langle H \rangle = 0
\end{cases}
$$

11.4 The Riemann Problem

An important problem of gas dynamics is the shock-tube problem. A tube is filled with a gas at rest in two different states separated by a diaphragm located at $x = x_0$. Interaction takes place after the membrane is broken at time $t = 0$. For a period of time, until reflections from the tube ends reach the test section, the problem can be treated as if the domain were infinite. The governing equations are the Euler equations ($g = 1$), equations (11.1), with initial conditions

$$
\begin{cases}
w_1 = \rho_l, \ w_2 = 0, \ w_3 = \rho_l E_l, \ x < x_0 \\
w_1 = \rho_r, \ w_2 = 0, \ w_3 = \rho_r E_r, \ x > x_0
\end{cases} ,
$$

where the subscripts l and r stand for left- and right-states. The initial conditions and the absence of boundary conditions makes for the *Riemann problem*. The solution is self-similar and can be expressed in terms of the ratio $\xi = (x - x_0)/t$, as $w(x, t) = W(\xi)$.

With the vector notation

$$\frac{\partial w}{\partial t} = -\frac{x - x_0}{t^2}\frac{dW}{d\xi},$$

$$\frac{\partial w}{\partial x} = \frac{1}{t}\frac{dW}{d\xi},$$

which leads to

$$-\frac{x - x_0}{t^2}\frac{dW}{d\xi} + \frac{\partial F}{\partial w}\frac{1}{t}\frac{dW}{d\xi} = \frac{1}{t}\left(\frac{\partial F}{\partial w} - \xi I\right).\frac{dW}{d\xi} = 0.$$

The trivial solution $W = $ const set aside, ξ is an eigenvalue of the Jacobian matrix and $dW/d\xi$ is a *right-eigenvector*. The solution is made of a combination of waves with increasing velocities from left to right.

11.5 A Box-Scheme for the Equations of Gas Dynamics

The classical centered schemes presented in Chapter 5 can be used to solve (11.1). The Lax scheme is dissipative enough to damp out oscillations, being first-order in space. The second-order accurate schemes, such as the Lax-Wendroff scheme and MacCormack scheme require some additional artificial viscosity.

An extension of the four-operator Murman scheme is presented that allows for the capture, without oscillations, of shocks and contact discontinuities. It is based on the compatibility relations (11.2) which form a system of three equations holding along the characteristic lines, that is equivalent to the original system, Reference [7].

Only three types of points are considered (a regular subsonic point and a regular supersonic point are treated in a similar fashion), depending on the number of characteristics converging toward a node i. The switches are based on the six discrete eigenvalues $\lambda_{i\pm\frac{1}{2}}^{(j)}$'s, where the mid-point values are calculated using the Roe averages, Reference [8] as for example:

$$\begin{cases} u_{i+\frac{1}{2}} = \dfrac{\sqrt{\rho_i}u_i + \sqrt{\rho_{i+1}}u_{i+1}}{\sqrt{\rho_i} + \sqrt{\rho_{i+1}}} \\[2mm] H_{i+\frac{1}{2}} = \dfrac{\sqrt{\rho_i}H_i + \sqrt{\rho_{i+1}}H_{i+1}}{\sqrt{\rho_i} + \sqrt{\rho_{i+1}}} \end{cases}, \quad a_{i+\frac{1}{2}}^2 = (\gamma - 1)\left(H_{i+\frac{1}{2}} - \frac{u_{i+\frac{1}{2}}^2}{2}\right).$$

The three cases are the following:

a) If three characteristics converge to point i, it is a *regular point*, and the three corresponding converging compatibility conditions are used. Say if $u_{i-\frac{1}{2}}^n - a_{i-\frac{1}{2}}^n < 0 < u_{i-\frac{1}{2}}^n$, and $u_{i+\frac{1}{2}}^n - a_{i+\frac{1}{2}}^n < 0 < u_{i+\frac{1}{2}}^n$, the scheme reads

$$
\begin{bmatrix} 0 \\ l^{(2)}_{i-\frac{1}{2}} \\ l^{(3)}_{i-\frac{1}{2}} \end{bmatrix} \cdot \left(\frac{w^{n+1}_i - w^n_i}{\Delta t} + \frac{f^n_i - f^n_{i-1}}{\Delta x} \right)
$$

$$
+ \begin{bmatrix} l^{(1)}_{i+\frac{1}{2}} \\ 0 \\ 0 \end{bmatrix} \cdot \left(\frac{w^{n+1}_i - w^n_i}{\Delta t} + \frac{f^n_{i+1} - f^n_i}{\Delta x} \right) = 0.
$$

b) If two or less characteristics converge to point i, it is a *sonic-point*, and one or more new equations are introduced corresponding to an expansion fan. If $u^n_{i-\frac{1}{2}} - a^n_{i-\frac{1}{2}} < 0 < u^n_{i-\frac{1}{2}}$, and $0 < u^n_{i+\frac{1}{2}} - a^n_{i+\frac{1}{2}}$, the scheme reads

$$
\begin{bmatrix} 0 \\ l^{(2)}_{i-\frac{1}{2}} \\ l^{(3)}_{i-\frac{1}{2}} \end{bmatrix} \cdot \left(\frac{w^{n+1}_i - w^n_i}{\Delta t} + \frac{f^n_i - f^n_{i-1}}{\Delta x} \right)
$$

$$
+ \begin{bmatrix} l^{(1)}_i \\ 0 \\ 0 \end{bmatrix} \cdot \left(\frac{w^{n+1}_i - w^n_i}{\Delta t} + \frac{\lambda^{(1)}_i \dfrac{w^n_{i+1} - w^n_{i-1}}{2\Delta x}}{1 + \dfrac{\Delta t}{\Delta x} \left(\lambda^{(1)}_{i+\frac{1}{2}} - \lambda^{(1)}_{i-\frac{1}{2}} \right)} \right) = 0.
$$

c) If four or more characteristics converge to point i, it is a *shock-point*. A conservative scheme consists in combining the box equations coming from the left and right boxes, using some vectors $\widetilde{l}^{(j)}_i$ as

$$
\begin{bmatrix} \widetilde{l}^{(1)}_i \\ \widetilde{l}^{(2)}_i \\ \widetilde{l}^{(3)}_i \end{bmatrix} \cdot \left(\frac{w^{n+1}_i - w^n_i}{\Delta t} + \frac{f^n_i - f^n_{i-1}}{\Delta x} + \frac{f^n_{i+1} - f^n_i}{\Delta x} \right) = 0.
$$

For an eigenvalue $\lambda^{(j)}$ which changes sign from $i - \frac{1}{2}$ to $i + \frac{1}{2}$, the scheme is similar to the mixed-type scheme for Burgers' equation, in particular as it concerns the shock- and sonic-point operators. Hence, this scheme computes the correct shock speed and prevents expansion shocks by forcing a smooth transition at sonic conditions. The time derivatives are obtained by solving a 3×3 system, whose matrix is composed of the left-eigenvectors.

The above scheme can be written in quasi-linear, but equivalent form, by making use of the property of the Roe averages. The following discrete relations are identically satisfied:

$$l_{i-\frac{1}{2}}^{(j)} \cdot \frac{f_i^n - f_{i-1}^n}{\Delta x} = \lambda_{i-\frac{1}{2}}^{(j)} l_{i-\frac{1}{2}}^{(j)} \cdot \frac{w_i^n - w_{i-1}^n}{\Delta x},$$

$$l_{i+\frac{1}{2}}^{(j)} \cdot \frac{f_{i+1}^n - f_i^n}{\Delta x} = \lambda_{i+\frac{1}{2}}^{(j)} l_{i+\frac{1}{2}}^{(j)} \cdot \frac{w_{i+1}^n - w_i^n}{\Delta x}.$$

For the sonic-point operator, the eigenvalue(s) and left-eigenvector(s) are evaluated at the nodal state w_i^n. This ensures that the scheme is not in conservation form, hence expansion shocks are excluded.

At the shock-point, four or more characteristics are converging to point i. The situation is identical to that of the Burgers' equation and is treated identically, by adding the space derivatives from the two surrounding boxes, thus producing a non-consistent but conservative scheme. The eigenvectors, denoted $\tilde{l}_i^{(j)}$, are combinations $\tilde{l}_i^{(j)} = \tilde{\theta} l_{i+\frac{1}{2}}^{(j)} + (1 - \tilde{\theta}) l_{i-\frac{1}{2}}^{(j)}$, that linearize the shock-point scheme. $\tilde{\theta}$ is obtained from

$$\tilde{\theta} l_{i+\frac{1}{2}}^{(j)} \cdot \left(\frac{\partial f_{i-\frac{1}{2}}^n}{\partial w} - \lambda_{i-\frac{1}{2}}^n I \right) \cdot \frac{w_i^n - w_{i-1}^n}{\Delta x}$$

$$+ (1 - \tilde{\theta}) l_{i-\frac{1}{2}}^{(j)} \cdot \left(\frac{\partial f_{i+\frac{1}{2}}^n}{\partial w} - \lambda_{i+\frac{1}{2}}^{(j)} I \right) \cdot \frac{w_{i+1}^n - w_i^n}{\Delta x} = 0.$$

With this choice, the scheme can be written, say for the compatibility relation CR^- as:

a) regular point

with $\lambda_{i-\frac{1}{2}}^{(1)} > 0$

$$l_{i-\frac{1}{2}}^{(1)} \cdot \left(\frac{w_i^{n+1} - w_i^n}{\Delta t} + \lambda_{i-\frac{1}{2}}^{(1)} \frac{w_i^n - w_{i-1}^n}{\Delta x} \right) = 0,$$

with $\lambda_{i+\frac{1}{2}}^{(1)} < 0$

$$l_{i+\frac{1}{2}}^{(1)} \cdot \left(\frac{w_i^{n+1} - w_i^n}{\Delta t} + \lambda_{i+\frac{1}{2}}^{(1)} \frac{w_{i+1}^n - w_i^n}{\Delta x} \right) = 0.$$

b) sonic-point

$$l_i^{(1)} \cdot \left(\frac{w_i^{n+1} - w_i^n}{\Delta t} + \frac{\lambda_i^{(1)} \frac{w_{i+1}^n - w_{i-1}^n}{2\Delta x}}{1 + \frac{\Delta t}{\Delta x} \left(\lambda_{i+\frac{1}{2}}^{(1)} - \lambda_{i-\frac{1}{2}}^{(1)} \right)} \right) = 0.$$

c) shock-point

$$\tilde{l}_i^{(1)} \cdot \left(\frac{w_i^{n+1} - w_i^n}{\Delta t} + \lambda_{i-\frac{1}{2}}^{(1)} \frac{w_i^n - w_{i-1}^n}{\Delta x} + \lambda_{i+\frac{1}{2}}^{(1)} \frac{w_{i+1}^n - w_i^n}{\Delta x} \right) = 0.$$

Problem: Using Von Neumann analysis, derive the stability conditions for the above schemes.

11.6 Some Results

The shock tube test case of Sod ($x_0 = 0.5$, $g = 1$, $\gamma = 1.4$), corresponding to the Riemann problem with initial conditions

$$\begin{cases} w_1 = 1, \ w_2 = 0, \ w_3 = 2.5, \ x < x_0 \\ w_1 = 0.125, \ w_2 = 0, \ w_3 = 0.25, \ x > x_0 \end{cases},$$

is simulated using $ix = 1001$ points for 500 steps with a CFL condition number equal to one. The results of the unsteady flow for density, velocity, energy and pressure are presented in Fig. 11.1 and are in good agreement with the exact solution. The solution is composed of, from left to right, a constant undisturbed left state, then a continuous expansion wave moving to the left, followed by a constant state, a contact discontinuity moving to the right, followed by a constant state, then a shock wave moving to the right in the undisturbed right state.

The flow of air ($\gamma = 1.4$) in a converging-diverging nozzle of equation $g(x) = 1 + (2x - 1)^2$ is marched to steady-state. The mesh has $ix = 101$ points. The variables are made dimensionless by the sonic conditions ρ^*, a^*, p^*.

The boundary conditions are at the entrance

$$\begin{cases} H_0 = \dfrac{\gamma p_0}{(\gamma - 1)\rho_0} + \dfrac{u_0^2}{2} = 3 \\ s_0 = C_v ln\left(\dfrac{\gamma p_0}{\rho_0^\gamma}\right) = 0 \end{cases},$$

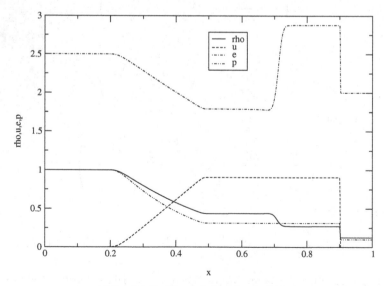

Fig. 11.1. Shock tube simulation

and at the exit

$$p_{\text{exit}} = 1.$$

The distributions of density, velocity, energy and pressure are shown in Fig. 11.2. Note the smooth sonic transition at the nozzle throat ($x_0 = 0.5$) and the abrupt recompression through a normal shock at $x = 0.82$.

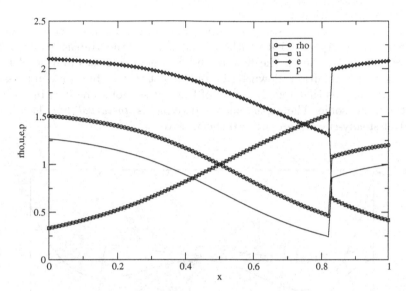

Fig. 11.2. Nozzle flow with shock wave

An interesting problem, that can be studied with this model, is the start-up of a supersonic wind tunnel having two throats, the first one of cross section A^* being fixed, the second one of variable section $A_2(\alpha) = 1 + (\alpha/4)$. The test section $A_{test} = 1.25A^*$, located at $x = 0.5$, is designed for an operating Mach number $M_{test} = 1.6$.

The geometry of the second nozzle is controlled by the parameter α as

$$\begin{cases} g(x) = \dfrac{9}{4}(2x - 1)^4 - \dfrac{3}{2}(2x - 1)^2 + \dfrac{5}{4}, & 0 \leq x \leq 0.5 \\[2mm] g(x) = (1 - \alpha)\left(\dfrac{9}{4}(2x - 1)^4 - \dfrac{3}{2}(2x - 1)^2\right) + \dfrac{5}{4}, & 0.5 \leq x \leq 1 \end{cases}$$

The uniform mesh system is defined by $x_i = (i - 1)\Delta x$, $\Delta x = 1/(ix - 1)$.

The start-up is simulated by the instantaneous opening of a valve, located between the first two points x_1 and x_2. The corresponding initial conditions are

$$\begin{cases} u_1 = 0 \\ \rho_1 = \left(\dfrac{\gamma+1}{2}\right)^{\frac{1}{\gamma-1}} \\ p_1 = \dfrac{\rho_1^\gamma}{\gamma} \end{cases} \qquad \begin{cases} u_i = 0 \\ \rho_i = \dfrac{2\gamma}{\gamma-1} p_{\text{exit}} \\ p_i = p_{\text{exit}} \end{cases}, \; i = 2, ..., ix \; .$$

In the application $p_{\text{exit}} = 0.9$.

In a first attempt to start the supersonic wind tunnel, the second throat area is set to $A_2 = 1.1\,A^*$ with $\alpha = 0.4$. After the transient shock has moved across the duct, The flow in the first nozzle chokes and accelerates to supersonic conditions downstream of the first throat, but a recompression shock terminates that region, ahead of the test section. The flow in the test section is subsonic. The supersonic wind tunnel is *unstarted*, Fig. 11.3. Note that at steady-state there are two shock waves.

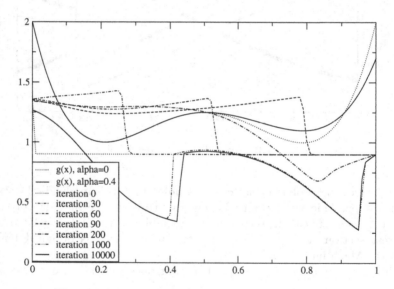

Fig. 11.3. Unstarted supersonic wind tunnel ($\alpha = 0.4$)

In order for the first shock to be "swallowed" by the second throat during the start-up, it is necessary that the second throat satisfy the inequality $A_2 \geq (A_2^*/A_1^*)[M_{\text{test}}]\,A^* = 1.1169\,A^*$, i.e. $\alpha \geq 0.4676$. If we increase the second throat to $A_2 = 1.15\,A^*$ with $\alpha = 0.6$, the flow goes through the transient motion of the shock, but now the test section is fully supersonic and shock free. The nominal Mach number $M_{\text{test}} = 1.6$ is obtained. There is a shock wave standing in the diffuser section of the second nozzle. The supersonic wind tunnel has been started, Fig. 11.4.

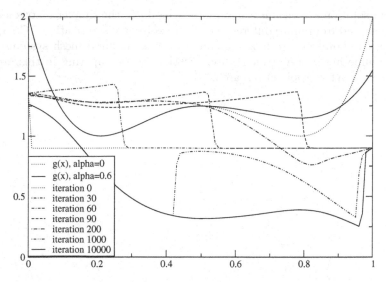

Fig. 11.4. Started supersonic wind tunnel ($\alpha = 0.6$)

These examples show how useful simulations can be in helping understand the physics of complex phenomena, such as shock tube flow or the start-up of a wind tunnel. In many cases, the steady-solution is well understood and available. But rarely are transient solutions known. Numerical simulation opens a window on this more challenging world.

11.7 The Bigger Picture

Other schemes are available, that have not been discussed in this book. Before closing this chapter we wish to mention a scheme that has been of comparable importance to the Murman-Cole scheme for the advance of research in the field of CFD: the Godunov scheme. It is based on the solution of the Riemann problem. Details can be found in text books, e.g. Reference [9].

Other discretization methods can be used to represent the continuum by a discrete set of points or functions. The finite element method (FEM) and the spectral method, use polynomial representation for the solution, the former on each element, the latter globally. The relationship between these different discretization approaches can be found in Reference [10].

The model problems have all been discretized on Cartesian mesh systems. More realistic geometries and more accurate boundary conditions require body fitted curvilinear mesh systems that can be treated, in a computational domain, with Cartesian meshes. There is a great advantage, when it is possible, to use a single Cartesian or structured mesh systems: high accuracy and efficient solution algorithms are available, thanks to the simpler matrix struc-ture obtained from finite difference or finite volume discretizations. Complex geometries, however, are more amenable to unstructured mesh systems. Fi-nite volume and finite element discretizations are competing in this arena, which is an active topic of research.

Appendix: A. Problems

A.1 Problem 1

Consider the following boundary value problem for the linear second-order ODE

$$\begin{cases} u''(x) = f(x), \ 0 \le x \le 1 \\ \qquad u(0) = 0 \\ \qquad u(1) = 0 \end{cases} \tag{A.1}$$

Two iterative methods are proposed to solve (A.1).

A.1.1 First Iterative Method

A uniform mesh is used with ix points $x_i = (i-1)\Delta x$, $\Delta x = 1/(ix-1)$. Let n be the iteration index for the "old" values (initial value when $n = 0$), and $n+1$ the index for the "new" values. The method reads:

$$\frac{u_{i+1}^n - 2u_i^{n+1} + u_{i-1}^n}{\Delta x^2} = f(x_i), \ i = 2, 3, ..., ix - 1, \tag{A.2}$$

$$\begin{cases} \qquad u_1^{n+1} = 0 \\ \qquad u_{ix}^{n+1} = 0 \\ u_i^0 = 0, \ i = 1, 2, ..., ix \end{cases} \tag{A.3}$$

Let $\delta u_i = u_i^{n+1} - u_i^n$ be the change of u_i between two iterations.

Time Evolution Equation Write method (A.2) in "delta form", that is with the δu_i in the left-hand side and the residual

$$R_i = \frac{u_{i+1}^n - 2u_i^n + u_{i-1}^n}{\Delta x^2} - f_i,$$

in the right-hand side, all the values being evaluated at the "old" time level.

If n is interpreted as a time index in a pseudo-time $\tau^n = n\Delta\tau$, such that $\delta u_i = \Delta\tau(\partial u_i^n / \partial\tau)$, perform a Taylor expansion of $u(x, \tau)$ to sufficient order and interpret the leading terms when $\Delta x, \Delta\tau \to 0$ as a time evolution PDE.

Equation Type What is the type of the PDE?

Consistency, Accuracy What is the relationship between $\Delta\tau$ and Δx required for the leading time derivative term to have a coefficient unity? Is this a classical result?

The method is consistent with the evolution PDE. What is the accuracy of the scheme in the transient? What is the accuracy at steady-state?

Exact Solution Check that, for $f(x) = -\pi^2 \sin \pi x$, the exact solution to the PDE, including initial and boundary conditions, is $u(x, \tau) = \left(1 - e^{-\pi^2 \tau}\right) \sin \pi x$.

A.1.2 Second Iterative Method

This method reads

$$\frac{u_{i+1}^n - 2u_i^{n+1} + u_{i-1}^{n+1}}{\Delta x^2} = f_i. \tag{A.4}$$

Equation (A.4) contains two "new" values.

Type of Scheme Sketch the computational molecule. Is this method implicit, i.e. does it require to solve a matrix equation? Hint: the points are swept with increasing values of i.

Evolution Equation Put the equation in delta form and find the time evolution equation governing the transient as $\Delta x, \Delta\tau \to 0$.

Consistency, Accuracy What is the condition for consistency if the coefficient of the unsteady term is unity? Is this method faster than the previous one? state the accuracy of the scheme during the transient and at steady-state. Is the steady-state solution different from that of the first method?

A.1.3 Implicit Scheme

Making the Scheme Implicit Propose an implicit scheme for problem (A.1) and sketch the computational molecule, using Fig. A.1.

Associated Matrix What type of matrix structure is associated with the implicit scheme? How would you solve the system of algebraic equations?

Convergence How many iterations will be needed for the implicit scheme to compute the steady solution? Will that number depend on the initial condition?

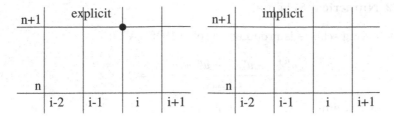

Fig. A.1. Sketch of computational molecule

A.2 Problem 2

Consider the initial-boundary value problem for the second-order PDE

$$\begin{cases} \dfrac{\partial^2 u}{\partial t \partial x} = f(x), \ x \geq 0, \ t \geq 0 \\[2mm] u(x,0) = g(x), \ x \geq 0 \\[2mm] u(0,t) = h(t), \ t \geq 0 \end{cases} \tag{A.5}$$

with $g(0) = h(0)$ (continuity condition).

A.2.1 Analytic Study

Equation Type What is the type of this PDE? Hint: introduce $v = \partial u / \partial t$, and transform the equation into a system of two first-order PDEs.

Characteristic Lines Find the characteristic directions and give the equations of the characteristic lines. Indicate with arrows, the direction of propagation of information along the characteristics. Hint: use the initial and boundary conditions.

General Solution Write the general solution to the homogeneous PDE, equation (A.5.1) with $f = 0$, in terms of two arbitrary functions $X(x)$ and $T(t)$.

Exact Solution Find the exact solution to (A.5) using two quadratures. First integrate in space from $x = 0$ to x, and apply the boundary condition as $(\partial u / \partial t)(0,t) = h'(t)$. Do not forget the arbitrary function.

Then integrate in time from $t = 0$ to t, and apply the initial condition. Do not forget the arbitrary function.

Example Let $u(x,t) = (1 + x + x^2)(1 + t)$ be the exact solution. Calculate f, g and h. Check your exact solution.

A.2.2 Numerical Scheme

The following scheme is proposed to solve PDE (A.5):

$$\frac{u_i^{n+1} - u_{i-1}^{n+1} - u_i^n + u_{i-1}^n}{\Delta t \Delta x} = f_{i-\frac{1}{2}}. \tag{A.6}$$

Type of Scheme Is (A.6) an implicit scheme? Hint: look at the boundary condition.

Consistency, Accuracy Perform the TE analysis of the above FDE. Define clearly the point about which you do the Taylor expansions. Is the scheme consistent? What is the accuracy in time and space?

Stability Introduce the complex modes $u_i^n = g^n e^{ii\alpha}$. Plug in the homogeneous equation and solve for the amplification factor. Is the scheme stable? Is there a stability condition?

Example Check if $u(x,t) = (1 + x + x^2)(1 + t)$ is also an exact solution to the FDE. You can use the TE result and compute the partial derivatives, or you can plug the exact solution in the FDE.

A.2.3 Numerical Application

A cartesian mesh with unit steps is defined by $x_i = (i - 1)$, $t^n = n$. The functions f, g and h are

$$\begin{cases} f = 1 \\ g = 1 + x \ . \\ h = 1 + t \end{cases}$$

Computational Molecule Sketch the computational molecule using Fig. A.2 and indicate how you compute u_i^{n+1} from the neighboring points.

Discrete Solution Compute a 5×5 block of data, i.e. $ix = 5$, $n = 0, 1, ..., 4$.

Discussion Compare the numerical results with the exact solution. Is the result in agreement with the TE analysis?

A.3 Problem 3

Consider the problem governed by the linear second-order PDE for the perturbation potential $\varphi(x,t)$

$$\frac{\partial^2 \varphi}{\partial t \partial x} + c\frac{\partial^2 \varphi}{\partial x^2} = 0, \ c > 0, \ x \geq 0, \ t \geq 0, \tag{A.7}$$

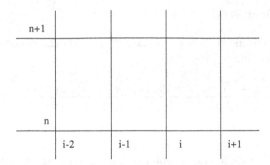

Fig. A.2. Sketch of computational molecule

subject to the initial and boundary conditions

$$
\begin{cases}
\varphi(x,0) = \begin{cases} \sin \pi x, \ 0 \le x \le 1 \\ \quad 0, \ x \ge 1 \end{cases} \\
\begin{cases} \varphi(0,t) = 0 \\ \dfrac{\partial \varphi}{\partial x}(0,t) = 0 \end{cases}, \ t \ge 0
\end{cases}
\tag{A.8}
$$

A.3.1 Analytic Study

Introduce a new unknown, $u = (\partial \varphi / \partial x)$, and reformulate (A.7) as a system of two first-order PDEs. Let $\phi(x,t) = $ const be the equation of a characteristic line in the (x,t) plane.

Equation Type Construct the characteristic matrix A and the characteristic form Q. Give the equation of the characteristic lines and conclude as to the type of the equation.

General Solution Give the general solution of the equation for $u(x,t)$. Integrate once more and find the general solution to the second-order PDE in terms of two arbitrary functions of a single argument. Check your result by plugging back into the PDE.

Exact Solution The domain $x \ge 0$, $t \ge 0$ can be divided into three regions from left to right.

Apply the boundary conditions at $x = 0$ and find the solution $\varphi(x,t)$ in the first region.

Find the value of $\varphi(x,t)$ on the boundary of 1–2.

Apply the initial condition at $t = 0$, $0 \le x \le 1$ and obtain the solution $\varphi(x,t)$ in region 2.

Proceed similarly for region 3.

A.3.2 Numerical Study

Define the uniform mesh $x_i = (i-1)\Delta x$, $t^n = n\Delta t$, $\Delta x = \text{const}$, $\Delta t = \text{const}$. The following scheme is proposed:

$$\frac{\varphi_i^{n+1} - \varphi_{i-1}^{n+1} - \varphi_i^n + \varphi_{i-1}^n}{\Delta t \Delta x} + c\frac{\varphi_i^n - 2\varphi_{i-1}^n + \varphi_{i-2}^n}{\Delta x^2} = 0.$$

Consistency, Accuracy Is this an implicit scheme? Hint: the solution is swept with increasing values of i.

Define the TE by choosing a point about which to expand in Taylor series and from the result conclude as to the consistency and accuracy of the above scheme. Carry the expansions to $O(\Delta t^2, \Delta t \Delta x, \Delta x^2)$.

Stability Let $\sigma = c(\Delta t/\Delta x) \geq 0$. Find the amplification factor g. Express the stability condition in terms of Δt.

A.3.3 Implicit Scheme

Devise an implicit scheme for this equation. Sketch the computational molecule in Fig. A.3.

Study the TE of the implicit scheme and conclude on its consistency and accuracy.

Find the amplification factor g for the implicit scheme. Is the stability unconditional?

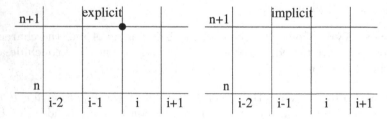

Fig. A.3. Sketch of the computational molecule

A.4 Problem 4

The nonlinear ODE

$$\frac{d}{dx}\left(\nu(u)\frac{du}{dx}\right) = 0, \; \nu(u) = \frac{u}{1+u}, \; u \geq 0, \; 0 \leq x \leq 1 \qquad (A.9)$$

with

$$\begin{cases} u(0) = 0 \\ u(1) = 1 \end{cases},$$

has a solution which satisfies the implicit equation

$$u - \ln(1 + u) = (1 - \ln 2)x.$$

Instead of solving this implicit equation, two finite difference methods are proposed to solve for u directly, using iterative procedures.

A.4.1 Explicit Scheme

This method consists in solving the following model problem:

$$\frac{\partial u}{\partial t} = \frac{\partial}{\partial x}\left(\nu(u)\frac{\partial u}{\partial x}\right),\ 0 \le x \le 1,\ t \ge 0,$$

subject to

$$\begin{cases} u(x,0) = 0,\ 0 \le x \le 1 \\ \begin{cases} u(0,t) = 0 \\ u(1,t) = 1 \end{cases},\ t \ge 0 \end{cases}.$$

Equation Type What is the type of this equation? Hint: introduce $v = \partial u/\partial x$ and study the corresponding first-order system in (u, v).

Characteristic Curves Find the characteristic lines of the PDE.

Characteristic Speed Taking into account the initial and boundary conditions, indicate with arrows the direction of propagation of information along the characteristics. What is the speed of propagation of the perturbations?

Consistency, Accuracy The discretization on a uniform mesh system, $x_i = (i-1)\Delta x$, $\Delta x = (1/ix - 1)$, $t^n = n\Delta t$, reads

$$\frac{u_i^{n+1} - u_i^n}{\Delta t} = \frac{\nu_{i+\frac{1}{2}}\dfrac{u_{i+1}^n - u_i^n}{\Delta x} - \nu_{i-\frac{1}{2}}\dfrac{u_i^n - u_{i-1}^n}{\Delta x}}{\Delta x},$$

where $\nu_{i\pm\frac{1}{2}} = \nu\left((u_{i\pm1}^n + u_i^n)/2\right)$.

The initial and boundary conditions are discretized in the obvious way (no discretization error). Define the TE, and study the consistency and accuracy of the explicit scheme by carrying the Taylor expansions so that the remainder is $O(\Delta t + \Delta x)^2$. What is the accuracy at steady-state? Hint: do the study for an arbitrary function $\nu(u)$.

Stability Assuming $\nu = $ const and letting $r = \nu(\Delta t/\Delta x^2)$, find the stability condition for Δt.

A.4.2 Semi-Implicit Scheme

This method corresponds to the model problem:

$$\frac{\partial^2 u}{\partial t \partial x} = \frac{\partial}{\partial x}\left(\nu(u)\frac{\partial u}{\partial x}\right), \ 0 \le x \le 1, \ t \ge 0,$$

subject to the same initial and boundary conditions.

Equation Type What is the type of this PDE?

Characteristic Curves What are the characteristic curves?

Characteristic Speed Taking into account the initial and boundary conditions, indicate with arrows the direction of propagation of information along the characteristics. Find the speed of propagation of the perturbations.

Consistency, Accuracy Perform the TE analysis of the semi-implicit scheme given below

$$\frac{u_i^{n+1} - u_{i-1}^{n+1} - u_i^n + u_{i-1}^n}{\Delta t \Delta x} = \frac{\nu_{i+\frac{1}{2}}\dfrac{u_{i+1}^n - u_i^n}{\Delta x} - \nu_{i-\frac{1}{2}}\dfrac{u_i^n - u_{i-1}^n}{\Delta x}}{\Delta x}.$$

Is the scheme consistent? What is the accuracy in time and space? What is the accuracy at steady-state? Is this an implicit scheme?

Stability Assume $\nu = $ const. Find the stability condition for Δt.

A.4.3 Implicit Scheme

Propose an implicit scheme to solve the steady-ODE, that is linear in the "new" values u^{n+1}. Sketch the computational molecule in Fig. A.4.

Give the coefficients of the matrix A, where $Au^{n+1} = b$.

How would you solve the algebraic system? Is it necessary to iterate?

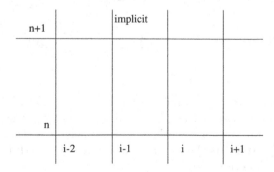

Fig. A.4. Sketch of computational molecule

A.5 Problem 5

Consider the first order PDE in "conservation form":

$$\frac{\partial u}{\partial t} - \frac{\partial}{\partial x}(xu) = 0, \quad -\infty < x < \infty, \; t \geq 0$$

$$u(x,0) = f(x).$$

A.5.1 Analytic Study

Equation Type What is the type of this PDE? Is it linear or nonlinear?

Characteristic Lines Find the slope of the characteristic $(dx/dt)_C$ and integrate to find the characteristic lines. Sketch the characteristic lines in the (x,t) plane.

Jump Conditions Study the jump conditions associated with the conservation form. Show that the jumps can only occur along characteristic lines (hint: find the slope of the line of discontinuity).

General Solution Write the compatibility relation, i.e. the governing equation for $u_c = u(t, x_c(t))$ where $x_c(t)$ is the equation of the characteristic line. Integrate the compatibility relation and find u_c.

Exact Solution Solve for $u(x,t)$, given the initial condition. Check your result.

A.5.2 Numerical Study

Discrete Conservation Form A uniform mesh is given by $x_i = i\Delta x$, $i = -ix, ..., -1, 0, 1, ..., ix$. The following scheme is used:

$$\frac{u_i^{n+1} - u_i^n}{\Delta t} - x_{i-\frac{1}{2}}\frac{u_i^n - u_{i-1}^n}{\Delta x} - \frac{u_{i-1}^n + u_i^n}{2} = 0, \; i < 0$$

$$\frac{u_i^{n+1} - u_i^n}{\Delta t} - x_{i+\frac{1}{2}}\frac{u_{i+1}^n - u_i^n}{\Delta x} - \frac{u_i^n + u_{i+1}^n}{2} = 0, \; i > 0.$$

Show that the discrete scheme can be cast in "conservation form" (hint: start from a conservation form and arrive at the above equations). Due to symmetries, only prove the result for $i < 0$.

Consistency, Accuracy Study the consistency and accuracy of this scheme. What is the accuracy of the transient solution? What is the accuracy at steady state?

Stability Study the stability of the scheme for the homogeneous equation. Show that the scheme has a stability requirement that may not be met for large values of $\mid x_{i-\frac{1}{2}} \mid$.

A.5.3 Implicit Scheme

The previous result pleads in favor of an implicit scheme. Devise an implicit scheme and study its stability. Is it unconditionally stable? Give the coefficients of the matrix for $i < 0$ and indicate how you can solve the linear system. Sketch the computational molecule in Figure A.5. Do not include the contribution of the source term in the computational molecule.

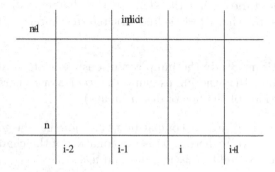

Fig. A.5. Sketch of computational molecule

A.6 Problem 6

Consider the following PDE:

$$\frac{\partial u}{\partial t} + \frac{\partial}{\partial x}\left(c(x)u\right) = f(x), \ -\infty < x < \infty, \ t \geq 0, \tag{A.10}$$

where $c(x)$ and $f(x)$ are continuous functions of x. This linear, non-homogeneous PDE completely determines the solution $u(x,t)$ when an initial condition is specified, i.e.

$$u(x,0) = u_0(x), \ -\infty < x < \infty.$$

A.6.1 Analytic Study

Equation Type Give the type of this equation and find the characteristic direction.

General Solution Let $v(x,t) = c(x)u(x,t)$, and $g(x) = c(x)f(x)$. Show that the PDE can be put in non-conservative form as

$$\frac{\partial v}{\partial t} + c(x)\frac{\partial v}{\partial x} = g(x). \tag{A.11}$$

One looks for the general solution in implicit form as

$$\Phi(x, t, v(x,t)) = 0. \tag{A.12}$$

Write two auxiliary relations by taking partial derivatives, $\partial/\partial t$, $\partial/\partial x$ of (A.12). Multiply (A.11) by $\partial\Phi/\partial v$ and show that

$$\frac{\partial\Phi}{\partial t} + c(x)\frac{\partial\Phi}{\partial x} + g(x)\frac{\partial\Phi}{\partial v} = 0. \tag{A.13}$$

(A.13) is a linear homogeneous PDE. Its general solution can be expressed in terms of first integrals

$$dt = \frac{dx}{c(x)} = \frac{dv}{g(x)}.$$

The solution can be written explicitly as a particular solution and an arbitrary function of a single argument:

$$v(x,t) = \int \frac{g(x)}{c(x)}dx + F\left(t - \int \frac{1}{c(x)}dx\right). \tag{A.14}$$

Check that (A.14) satisfies (A.11).

Application Let $c(x) = 2x - 1$ and $f(x) = 4(2x - 1) = 4c(x)$.

Find the equation of the characteristic. Draw several characteristic lines in the (x, t) plane.

These functions for $c(x)$ and $f(x)$ will be used in the rest of the problem.

Remark: The integration of the steady-state solution $(d/dx)((2x - 1)u) = 4(2x - 1)$ as an ODE from $x = 0$ to $x = 1$ with the initial condition $u(0) = -1$ is very difficult because the coefficient $c(x) = 2x - 1$ vanishes in the interval. The time evolution equation circumvents this difficulty. Although the equation is linear, it shares many features with the nonlinear transonic nozzle flow model: an initial condition is needed, but no boundary condition, as is the case for the sonic expansion.

A.6.2 Numerical Scheme

A uniform mesh is defined, $x_i = (i - 1)\Delta x$, $\Delta x = (1/ix - 1)$, and Δt is the time step. Let $c_{i\pm\frac{1}{2}} = 2x_{i\pm\frac{1}{2}} - 1$ where $x_{i\pm\frac{1}{2}} = x_i \pm \frac{\Delta x}{2}$. Let $u_{i\pm\frac{1}{2}} = (u_{i\pm1} + u_i)/2$. The following scheme is proposed:

i) $c_{i-\frac{1}{2}} > 0$ and $c_{i+\frac{1}{2}} > 0$

$$\frac{u_i^{n+1} - u_i^n}{\Delta t} + c_{i-\frac{1}{2}}\frac{u_i^n - u_{i-1}^n}{\Delta x} = 4c_{i-\frac{1}{2}} - 2u_{i-\frac{1}{2}}^n,$$

ii) $c_{i-\frac{1}{2}} > 0$ and $c_{i+\frac{1}{2}} < 0$

$$\frac{u_i^{n+1} - u_i^n}{\Delta t} + c_{i-\frac{1}{2}}\frac{u_i^n - u_{i-1}^n}{\Delta x} + c_{i+\frac{1}{2}}\frac{u_{i+1}^n - u_i^n}{\Delta x}$$

$$= 4\left(c_{i-\frac{1}{2}} + c_{i+\frac{1}{2}}\right) - 2\left(u_{i-\frac{1}{2}}^n + u_{i+\frac{1}{2}}^n\right),$$

iii) $c_{i-\frac{1}{2}} < 0$ and $c_{i+\frac{1}{2}} > 0$

$$\frac{u_i^{n+1} - u_i^n}{\Delta t} + c_i\frac{u_{i+1}^n - u_{i-1}^n}{2\Delta x} = 4c_i^n - 2u_i^n,$$

iv) $c_{i-\frac{1}{2}} < 0$ and $c_{i+\frac{1}{2}} < 0$

$$\frac{u_i^{n+1} - u_i^n}{\Delta t} + c_{i+\frac{1}{2}}\frac{u_{i+1}^n - u_i^n}{\Delta x} = 4c_{i+\frac{1}{2}} - 2u_{i+\frac{1}{2}}^n.$$

Consistency, Accuracy Draw the computational molecule in each case, using Figure A.6. Study the consistency and accuracy at steady-state of all four schemes using TE analysis with remainder $O(\Delta x^2)$. Note that there are no errors on $c_{i\pm\frac{1}{2}}$ because they are linear functions of x.

The steady-state solution is $u_{ss}(x) = 2x - 1$. Check that it is the solution of the finite difference schemes.

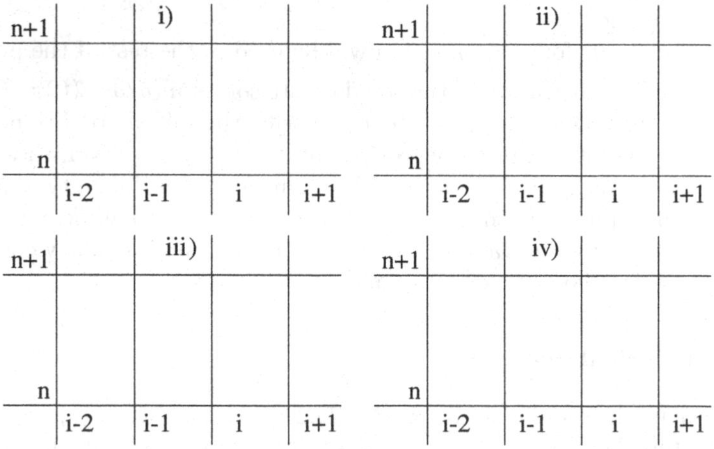

Fig. A.6. Sketch of computational molecules

Stability Study the stability of the time marching procedure and define the maximum time step allowable for each scheme, using Von Neumann analysis. Note that the equations are linear, but are made homogeneous by setting the right-hand sides to zero.

A.6.3 Implicit Scheme

Devise an implicit scheme based on the four-operator schemes above, and draw the corresponding computational molecules. No boundary condition needs to be specified for this problem.

Build the matrix and right-hand sides associated with your implicit scheme in the case $ix = 4$. Is the matrix diagonally dominant? Is there a restriction on the time step?

What solution procedure can you use to solve the linear system of equations for large values of ix?

A.7 Problem 7

A.7.1 The MacCormack Scheme

Let $u(x, t)$ be the unknown scalar function and $F(u)$ an arbitrary flux function. The two-steps MacCormack scheme reads:

$$\overline{u_i^{n+1}} = u_i^n - \frac{\Delta t}{\Delta x}\left(F_{i+1}^n - F_i^n\right),\tag{A.15}$$

$$u_i^{n+1} = \frac{1}{2}\left(u_i^n + \overline{u_i^{n+1}} - \frac{\Delta t}{\Delta x}\left(\overline{F_i^{n+1}} - \overline{F_{i-1}^{n+1}}\right)\right).\tag{A.16}$$

Consistency, Accuracy Rewrite the scheme as a FDE for the PDE

$$\frac{\partial u}{\partial t} + \frac{\partial F}{\partial x} = 0.$$

This is done by eliminating the intermediate value $\overline{u^{n+1}}$ from the scheme, equation (A.16) and dividing by Δt.

Then, expand each FD quotient in the TE in Taylor series to sufficient order, so that the final remainder is $O(\Delta t + \Delta x)^2$.

Show that the scheme is second-order accurate in space and time.

Stability Let $F(u) = cu$, $c = $ const of arbitrary sign. Let $\sigma = c(\Delta t/\Delta x)$ be the Courant number. Prove that the scheme is stable if $|\sigma| \leq 1$.

A.7.2 Exact Solution to Burgers' Equation (Inviscid)

General Solution Prove that the general solution of Burgers' equation

$$\frac{\partial u}{\partial t} + \frac{\partial}{\partial x}\left(\frac{u^2}{2}\right) = 0, \tag{A.17}$$

is given implicitly by

$$u(x,t) = G\left(x - tu(x,t)\right),$$

where $G(\xi)$ is an arbitrary function of a single argument $\xi = x - tu(x,t)$.

Initial Value Problem Given the initial condition

$$u(x,0) = \begin{cases} \sqrt{-4x}, & x \leq 0 \\ 0, & x \geq 0 \end{cases},$$

find the unknown function $G(\xi)$ for all values of its argument.

Give the equation of the characteristic lines passing through a point $(x_0, 0)$ in the form $\phi(x,t;x_0) = 0$.

Sketch, in the (x,t) plane, some characteristic lines and show the necessity of introducing a shock wave.

Shock Curve Find the envelop of the characteristics by solving the system for x and t given by

$$\begin{cases} \phi(x,t;x_0) = 0 \\ \dfrac{\partial \phi}{\partial x_0}(x,t;x_0) = 0 \end{cases}.$$

Show that the envelop starts at $(0,0)$, hence the shock will starts there also.

Write the jump conditions and show that they are not satisfied on the envelop.

Look for a shock curve S of the form $x = at^2$, $a = const > 0$. Find a.

Sketch the resulting solution in the (x,t) plane.

A.7.3 Numerical Study

Find the exact solution to the following nonlinear second-order ODE:

$$\frac{d^2}{dx^2}\left(\frac{u^2}{2}\right) = 0,$$

subject to the boundary conditions

$$\begin{cases} u(0) = 0 \\ u(1) = 1 \end{cases}.$$

Discrete Solution A uniform mesh is constructed as $x_i = (i-1)\Delta x$, $\Delta x = 1/(ix-1)$. The discrete solution satisfies

$$\frac{u_{i+\frac{1}{2}}\dfrac{u_{i+1}-u_i}{\Delta x} - u_{i-\frac{1}{2}}\dfrac{u_i-u_{i-1}}{\Delta x}}{\Delta x} = 0, \; i = 2, ..., ix-1$$

where $u_{i\pm\frac{1}{2}} = \frac{u_{i\pm1}+u_i}{2}$.

The boundary conditions are $u_1 = 0$, $u_{ix} = 1$.

Prove that the exact solution satisfies this FDE.

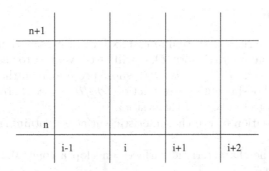

Fig. A.7. Sketch of computational molecule

Iterative Method The iterative procedure to solve the above equations reads:

$$\frac{u_{i+1}^{n+1} - u_i^{n+1} - u_{i+1}^n + u_i^n}{\Delta x^2} + \frac{u_{i+\frac{1}{2}}^n\dfrac{u_{i+1}^n-u_i^n}{\Delta x} - u_{i-\frac{1}{2}}^n\dfrac{u_i^n-u_{i-1}^n}{\Delta x}}{\Delta x} = 0,$$

$$i = ix-1, ..., 2.$$

Draw the computational molecule using Fig. A.7. Note that the domain is swept in reverse direction, with decreasing values of the index i.

Interpret this iterative algorithm as a time evolution PDE. Hint: introduce a time variable such that $t^n = n\Delta t$ and assume that $\Delta x, \Delta t \to 0$.

What choice of Δt will produce a meaningful PDE with a coefficient unity for the time derivative?

Find the type of the PDE and the characteristic curves.

Sketch in the (x, t) plane the characteristic lines and indicate with arrows the direction of propagation of the information on them (assume that $u > 0$ everywhere).

Solution to the Evolution Equation Given the initial condition

$$\begin{cases} u_1^0 = 0 \\ u_i^0 = 1, \; i = 2, ..., ix \end{cases},$$

find the solution to the time evolution PDE until $t = 1$. Hint: integrate the PDE once in x and use known results.

A.8 Problem 8

Consider the linear system of two first-order PDEs for $u(x, y, t)$ and $v(x, y, t)$:

$$\begin{cases} \dfrac{\partial u}{\partial t} + \dfrac{\partial u}{\partial x} + \dfrac{\partial v}{\partial y} = r_1 = 0 \\[2mm] \dfrac{\partial v}{\partial t} - \dfrac{\partial v}{\partial x} + \dfrac{\partial u}{\partial y} = r_2 = 0 \end{cases} \quad -\infty < x, y < \infty, \ t \geq 0. \qquad (A.18)$$

A.8.1 Analytic Study

System Type Write the system in matrix form. Find the characteristic matrix A and the characteristic form Q. It will be convenient to use a normal vector $\nabla \phi = ((\partial \phi / \partial x), (\partial \phi / \partial y), (\partial \phi / \partial t))$ whose projection on the (x, y) plane is normalized by choosing without restriction $(\partial \phi / \partial x) = \cos \theta$, $(\partial \phi / \partial y) = \sin \theta$.

Conclude as to the type of the system.

Give the equation of the characteristic surfaces containing the origin, in terms of θ.

Show that the characteristic surfaces envelop a cone. What is the half-angle of the cone?

Compatibility Relations Find the two compatibility relations for system (A.18) in terms of r_1, r_2 and θ. Are these relations independent for all θ? Introduce the half-angle $\theta/2$ and use the identities $(\sin \theta = 2 \sin(\theta/2) \cos(\theta/2)$, $\cos \theta = 1 - 2 \sin^2(\theta/2)$, to simplify the compatibility relations and ensure their independence.

Jump Conditions Find the jump conditions associated with the PDEs. Show that the jumps can only occur along characteristic surfaces.

Exact Solutions to Two Initial Value Problems Use the compatibility relations to find u and v if for $t = 0$:

$$\begin{cases} i) \ u = v = 0, \ x < 0; \ u = v = 1, \ x > 0; \ \forall y \\ ii) \ u = v = 1, \ y < 0; \ u = v = 0, \ y > 0; \ \forall x \end{cases}.$$

Hint: in the first case, the problem is 1-D, choose $\theta = 0$.

In each case make a sketch of the (x, t) plane and (y, t) plane, and indicate the regions where the solution is constant. Check in each case that the jump relations are satisfied.

A.8.2 Numerical Study

The system is discretized on a staggered mesh, with values of u and v at the mid-edges, as shown in Fig. A.8. The first equation is discretized on the solid box, the second on the dotted box as:

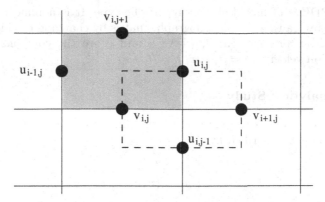

Fig. A.8. Staggered mesh configuration and computational molecule

$$\begin{cases} \dfrac{u_{i,j}^{n+1} - u_{i,j}^{n}}{\Delta t} + \dfrac{u_{i,j}^{n} - u_{i-1,j}^{n}}{\Delta x} + \dfrac{v_{i,j+1}^{n+1} - v_{i,j}^{n+1}}{\Delta y} = 0 \\[4mm] \dfrac{v_{i,j}^{n+1} - v_{i,j}^{n}}{\Delta t} - \dfrac{v_{i+1,j}^{n} - v_{i,j}^{n}}{\Delta x} + \dfrac{u_{i,j}^{n+1} - u_{i,j-1}^{n+1}}{\Delta y} = 0 \end{cases}.$$

Consistency, Accuracy Using the TE analysis, derive the consistency and accuracy of the scheme. Hint: expand about the center of each box. What is the accuracy at steady-state?

Stability Introduce $\sigma_x = \Delta t/\Delta x$, $\sigma_y = \Delta t/\Delta y$, and let

$$\begin{cases} u_{i,j}^{n} = U^{n} e^{ii\alpha} e^{ij\beta} \\ v_{i,j}^{n} = V^{n} e^{ii\alpha} e^{ij\beta} \end{cases}.$$

Find the 2×2 amplification matrix G such that

$$\begin{bmatrix} U^{n+1} \\ V^{n+1} \end{bmatrix} = G. \begin{bmatrix} U^{n} \\ V^{n} \end{bmatrix}.$$

Find the stability condition. Hint: find the eigenvalues $\lambda_{1,2}$ of G and satisfy $|\lambda_{1,2}| \leq 1$.

A.9 Problem 9

Consider the transonic small disturbance equation (TSD)

$$-(\gamma + 1)\frac{\partial \varphi}{\partial x}\frac{\partial^2 \varphi}{\partial x^2} + \frac{\partial^2 \varphi}{\partial y^2} = 0. \tag{A.19}$$

This PDE is of mixed elliptic-hyperbolic type. It is nonlinear. It can be linearized by transfer to the hodograph, that is by changing the independent variables from (x, y) to (u, v), where u and v are the components of the perturbation velocity, $u = \partial\varphi/\partial x$, $v = \partial\varphi/\partial y$.

A.9.1 Analytical Study

Transfer to the Hodograph The mapping from the hodograph to the physical plane is described by

$$\begin{cases} x = x(u, v) \\ y = y(u, v) \end{cases}.$$

Derive the differential relations for dx and dy in terms of du and dv.

Solve this linear system for (du, dv), assuming that the Jacobian of the transformation is not singular, i.e.

$$J = \frac{\partial x}{\partial u}\frac{\partial y}{\partial v} - \frac{\partial x}{\partial v}\frac{\partial y}{\partial u} \neq 0, \infty.$$

From these, find $\partial u/\partial x$ and $\partial v/\partial y$ in terms of the partial derivatives of the mapping.

Introduce the Legendre potential $\Phi(u, v) = xu + yv - \varphi(x, y)$ and show that $\partial\Phi/\partial u = x$, $\partial\Phi/\partial v = y$.

Transform equation (A.19) into a first-order PDE for x and y, using the identities $\partial\varphi/\partial x = u$, $\partial^2\varphi/\partial x^2 = \partial u/\partial x$, $\partial^2\varphi/\partial y^2 = \partial v/\partial y$.

Write a second first-order PDE for x and y, by taking cross-derivatives $(\partial^2\Phi/\partial u\partial v)$.

Finally, obtain a second-order PDE for Φ by eliminating x and y using the last two relations, and show that the transformed potential equation reads

$$\frac{\partial^2\Phi}{\partial u^2} - (\gamma + 1)\, u\frac{\partial^2\Phi}{\partial v^2} = 0. \tag{A.20}$$

This is Tricomi's equation. It is linear in the unknown Φ.

Equation Type Study Tricomi's equation type. It will be convenient to use the equivalent system of two first-order PDEs for x and y derived previously.

What is the equation of the sonic line?

Find the slopes of the characteristics in the hyperbolic subdomain. Integrate to find the equation of the characteristics and sketch them in the (u, v) plane.

A.9.2 Numerical Study

The computational domain in the (u, v) plane is shown in Fig. A.9. A cartesian mesh is introduced as

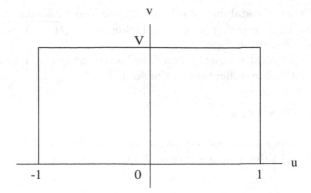

Fig. A.9. Computational domain for Tricomi's equation

$$
\begin{cases}
u_i = -1 + (i-1)\Delta x, \ \Delta x = \dfrac{2}{ix-1}, \ i = 1, ..., ix \\[2mm]
v_j = (j-1)\Delta v, \ \Delta v = \dfrac{V}{jx-1}, \ j = 1, ..., jx
\end{cases}
$$

The following scheme is proposed:
i) $u_{i-1} > 0$, $u_i > 0$ (supersonic point)

$$
\frac{\Phi^{n+1}_{i,j} - 2\Phi^{n+1}_{i-1,j} + \Phi^{n+1}_{i-2,j}}{\Delta u^2} - (\gamma+1)u_{i-1}\frac{\Phi^{n+1}_{i-1,j+1} - 2\Phi^{n+1}_{i-1,j} + \Phi^{n+1}_{i-1,j-1}}{\Delta v^2} = 0,
$$

ii) $u_{i-1} \le 0$, $u_i > 0$ (sonic point)

$$
\frac{\Phi^{n+1}_{i,j} - 2\Phi^{n+1}_{i-1,j} + \Phi^{n+1}_{i-2,j}}{\Delta u^2} - (\gamma+1)u_i\frac{\Phi^{n+1}_{i-1,j+1} - 2\Phi^{n+1}_{i-1,j} + \Phi^{n+1}_{i-1,j-1}}{\Delta v^2} = 0,
$$

iii) $u_i \le 0$ (subsonic point)

$$
\frac{\Phi^{n}_{i+1,j} - 2\widetilde{\Phi}_{i,j} + \Phi^{n+1}_{i-1,j}}{\Delta u^2} - (\gamma+1)u_i\frac{\Phi^{n}_{i,j+1} - 2\widetilde{\Phi}_{i,j} + \Phi^{n+1}_{i,j-1}}{\Delta v^2} = 0,
$$

Where the tilde represents a predicted value to be over-relaxed according to $\Phi^{n+1} = \Phi^n + \omega(\widetilde{\Phi} - \Phi^n)$. ω is the relaxation factor, $1 \le \omega < 2$.

Consistency, Accuracy Define the TE and perform the Taylor expansions for each of the schemes at convergence (drop the upper index n). Are the schemes consistent? What is their accuracy?

Iterative Method Describe the iterative method in each case. Can you relate it to known iterative methods?

Stability Study the stability of scheme i). Use Von Neumann analysis after linearization. Hint: use $\Phi_{i,j} = g^i e^{ij\beta}$. Define $\sigma = \sqrt{(\gamma+1)u_{i-1}}(\Delta u/\Delta v)$. Find the CFL condition for σ.

Where is the most restrictive condition obtained in the domain?

Do you anticipate difficulties at the sonic line?

A.9.3 Implicit Scheme

Propose a SLOR scheme for Tricomi's equation. Sketch the computational molecules for all three schemes in Fig. A.10.

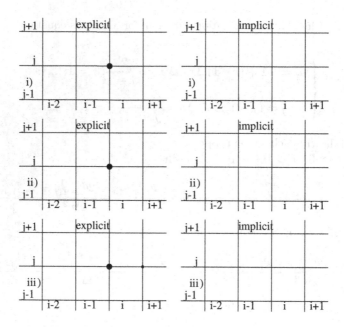

Fig. A.10. Sketch of computational molecule

What structure will the matrix have? Is the matrix diagonally dominant? How will you solve the algebraic system?

A.10 Problem 10

Consider the low-frequency transonic small disturbance equation (TSD):

$$\frac{\partial^2 \varphi}{\partial t \partial x} + \frac{\partial}{\partial x}\left(\frac{1}{2}\left(\frac{\partial \varphi}{\partial x}\right)^2\right) = 0, \tag{A.21}$$

with the following initial conditions:

$$
\begin{cases}
\varphi(x,0) = x, \ 0 \leq x \leq \tfrac{1}{5} \\[2mm]
\varphi(x,0) = \tfrac{1}{5}, \ \tfrac{1}{5} \leq x \leq \tfrac{1}{2} \\[2mm]
\varphi(x,0) = \dfrac{1}{5} - \dfrac{1}{2}\left(x - \dfrac{1}{2}\right), \ \dfrac{1}{2} \leq x \leq 1
\end{cases}
,
$$

and boundary conditions:

$$
\begin{cases}
\varphi(0,t) = 0 \\[2mm]
\dfrac{\partial \varphi}{\partial x}(0,t) = 1 \quad , \ t \geq 0 \ . \\[2mm]
\dfrac{\partial \varphi}{\partial x}(1,t) = -\dfrac{1}{2}
\end{cases}
$$

A.10.1 Analytic Study

Equation Type Set equation (A.21) in non-conservative form and transform it into a system of two first-order PDEs upon introducing $u = \partial\varphi/\partial x$.

Let $\phi(x,t) = \text{const}$ be the equation of a curve, find the type of the system. Find the characteristic curves. Sketch the two cases: $u \geq 0$, and $u \leq 0$.

Jump Conditions Find the jump conditions associated with the system (u, φ). Find the slope of the jump line.

Exact Solution for $u(x,t)$ Using the initial and boundary conditions, sketch the different regions in the (x,t) plane and give the exact solution $u(x,t)$ in each sub-region with the inequalities defining the boundaries of each sub-region.

Exact Solution for $\varphi(x,t)$ Find the exact solution for φ.

A.10.2 Numerical Scheme

A mesh is defined as $x_i = (i-1)\Delta x$, $\Delta x = (1/ix - 1)$.

The following switch is introduced:

$$
u_i = \frac{\varphi_{i+1}^n - \varphi_{i-1}^n}{2\Delta x}.
$$

A mixed-scheme is defined as:

i) $u_{i-1} > 0$ and $u_i > 0$ (supersonic point)

$$
\frac{\varphi_i^{n+1} - \varphi_{i-1}^{n+1} - \varphi_i^n + \varphi_{i-1}^n}{\Delta t \Delta x} + u_{i-1}\frac{\varphi_i^n - 2\varphi_{i-1}^n + \varphi_{i-2}^n}{\Delta x^2} = 0,
$$

ii) $u_{i-1} > 0$ and $u_i < 0$ (shock point)

$$\frac{\varphi_i^{n+1} - \varphi_{i-1}^{n+1} - \varphi_i^n + \varphi_{i-1}^n}{\Delta t \Delta x} + u_{i-1} \frac{\varphi_i^n - 2\varphi_{i-1}^n + \varphi_{i-2}^n}{\Delta x^2}$$

$$+ u_i \frac{\varphi_{i+1}^n - 2\varphi_i^n + \varphi_{i-1}^n}{\Delta x^2} = 0,$$

iii) $u_{i-1} < 0$ and $u_i > 0$ (sonic point)
iv) $u_{i-1} > 0$ and $u_i < 0$ (supersonic point)

$$\frac{\varphi_i^{n+1} - \varphi_{i-1}^{n+1} - \varphi_i^n + \varphi_{i-1}^n}{\Delta t \Delta x} + u_i \frac{\varphi_{i+1}^n - 2\varphi_i^n + \varphi_{i-1}^n}{\Delta x^2} = 0.$$

The sonic point operator has not been defined and is not used in this problem.

Consistency and Accuracy Study the consistency of the above schemes and state their accuracy, using the TE analysis. Note that in all cases the first term is the same. What is the accuracy at steady-state?

Stability Linearize the difference equations as suggested by the above formulation, using the Courant numbers

$$\sigma^- = u_{i-1} \frac{\Delta t}{\Delta x}, \ \sigma^+ = -u_i \frac{\Delta t}{\Delta x},$$

and study the stability of the various schemes. Find the stability condition for Δt. Note that the stability condition for the shock point encompasses that of the other schemes.

A.10.3 Implicit Scheme

Making the Scheme Implicit How would you make the above scheme implicit?

Sketch the computational molecule using Fig. A.11.

Consistency, Accuracy What is the expected accuracy of the implicit schemes i), ii) and iv)?

Matrix Structure What is the structure of the matrix of the implicit scheme? How would you solve the algebraic system?

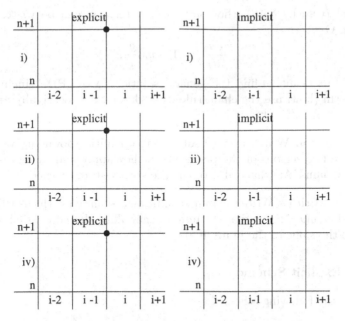

Fig. A.11. Sketch of the computational molecule

A.11 Problem 11

Consider the following PDE, IC and BCs for $u(x,t)$:

$$\frac{\partial u}{\partial t} + \frac{\partial u}{\partial x} = (1-u)\frac{g'}{g},$$

$$u(x,0) = 0,\ 0 \leq x \leq 1,$$

$$u(0,t) = 0,\ t \geq 0.$$

$g(x)$ represents the area of the supersonic nozzle diffuser, where $g(0) = 1$ and $g(x) \geq 1,\ 0 \leq x \leq 1$.

A.11.1 Analysis of the Problem

Equation Type Is the PDE linear? Find the type of the equation. Does the type depend on $g(x)$?

Characteristic Lines Find the slope and the equation of the characteristic line.

Compatibility Relation Give the compatibility relation. Show that if $u_c[x, t_c(x)]$ represents the value of u along the characteristic of equation

$t = t_c(x)$ (Cauchy data), the compatibility relation can be written as the following ODE

$$\frac{du_c}{dx} = (1 - u_c)\frac{g'}{g},$$

and use this result to find the general solution to the PDE, including the source term (hint: use the chain rule). Check your signs carefully and verify your result by differentiation.

Exact Solution Write the exact solution to the initial/boundary value problem in the two regions of the plane where they apply and make a sketch of the subdomains. At what value of t is the steady-state reached?

Jump Conditions Write the jump condition associated with the PDE. Does it depend on $g(x)$? Show that the jumps occur along characteristic lines (hint: compute the slope of the jump line).

A.11.2 Explicit Scheme

Consider the following scheme

$$\frac{u_i^{n+1} - u_i^n}{\Delta t} + \frac{u_i^n - u_{i-1}^n}{\Delta x} - (1 - u_{i-1}^n)\frac{g_i - g_{i-1}}{\Delta x\, g_i} = 0.$$

Sketch the computational molecule. Indicate on your sketch the characteristic C of the PDE and the numerical characteristic N. Show with an arrow the direction of propagation of the information on the characteristics. What is the speed of propagation of the information on the C and N characteristics?

Consistency, Accuracy Choose a convenient point and expand with care the truncation error in Taylor series to order $O(\Delta t^2, \Delta x^2)$, i.e. get the first-order terms with full details and collect the second- and higher-order terms in the "garbage". Note that the source term is made of three parts. Is the scheme consistent? What is its accuracy during the transient? What is its accuracy at steady-state?

Substitute the exact steady solution found in 1.4 into the steady equation. Conclude.

Stability Study the stability of the scheme. If there is a stability condition, give the maximum time step allowable.

A.11.3 Implicit Scheme

The following implicit scheme is proposed:

$$\frac{u_i^{n+1} - u_i^n}{\Delta t} + \frac{u_i^{n+1} - u_{i-1}^{n+1}}{\Delta x} - (1 - u_{i-1}^{n+1})\frac{g_i - g_{i-1}}{\Delta x\, g_i} = 0.$$

Sketch the computational molecule. Indicate on your sketch the characteristic C of the PDE and the numerical characteristic N. Show with an arrow the direction of propagation of the information on the characteristics. What is the speed of propagation of the information on the C and N characteristics?

Consistency, Accuracy Study the truncation error and conclude as to the consistency and accuracy of the scheme (hint: you can reuse some of your previous Taylor expansions). What is the accuracy at steady-state?

Stability Study the stability of the implicit scheme. Conclude.

Solution Procedure Show the structure of the matrix of the iterative method (conditioning matrix) associated with this implicit scheme (hint: use the delta form where $\delta u_i = u_i^{n+1} - u_i^n$). If the matrix fits in a tridiagonal matrix structure, find the coefficients p_i, q_i, r_i and s_i. Is there a need for a solver? Indicate how you would compute the solution at the new time step.

Appendix: B. Solutions to Problems

B.1 Solution to Problem 1

B.1.1 First Iterative Method

Time Evolution Equation In delta form, the numerical scheme reads:

$$\frac{2}{\Delta x^2} \delta u_i = \frac{u_{i+1}^n - 2u_i^n + u_{i-1}^n}{\Delta x^2} - f_i = R_i.$$

This is Jacobi's method. Performing a Taylor expansion about (i, n) yields

$$2\frac{\Delta \tau}{\Delta x^2}\frac{\partial u_i^n}{\partial \tau} + O\left(\frac{\Delta \tau^2}{\Delta x^2}\right) = \frac{\partial^2 u_i^n}{\partial x^2} + O(\Delta x^2) - f_i.$$

As $\Delta x, \Delta \tau \to 0$, the non trivial evolution equation corresponding to this iterative method is

$$\frac{\partial u}{\partial \tau} = \frac{\partial^2 u}{\partial x^2} - f. \tag{B.1}$$

Equation Type We recognize equation (B.1) as the heat equation with a source term. It is parabolic.

Consistency, Accuracy Equation (B.1) corresponds to the choice $\Delta \tau = \Delta x^2/2$. With this choice, the above scheme corresponds exactly to the explicit scheme for the heat equation seen in Chap. 6, with $\nu = 1$ and x replacing y. Hence, the classical stability condition for the heat equation using an explicit scheme is recovered. The TE is also identical:

$$\epsilon_i^n = O\left(\frac{\Delta \tau^2}{\Delta x^2}, \Delta x^2\right) = O(\Delta \tau, \Delta x^2).$$

In the transient, the scheme is first-order accurate in τ and second-order accurate in x.

At steady-state, the scheme is second-order accurate in x.

Exact Solution $u(x, \tau) = \left(1 - e^{-\pi^2 \tau}\right) \sin \pi x$ is the exact solution to the continuous problem (PDE+IC+BC's) since

$$\frac{\partial u}{\partial \tau} = \pi^2 e^{-\pi^2 \tau} \sin \pi x, \quad \frac{\partial^2 u}{\partial x^2} = -\pi^2 \left(1 - e^{-\pi^2 \tau}\right) \sin \pi x, \quad f(x) = -\pi^2 \sin \pi x.$$

B.1.2 Second Iterative Method

Type of Scheme The computational molecule is shown in Fig. B.1.

This is not an implicit scheme. This is Gauss-Seidel method. Starting from $i = 2$, the boundary condition gives u_1^{n+1}, and the only unknown is u_2^{n+1}. By induction, the only unknown in the equation at point i is u_i^{n+1}, which can be solved for explicitly.

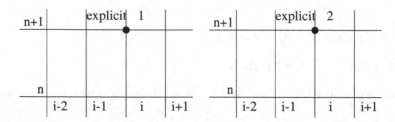

Fig. B.1. Sketch of computational molecule

Evolution Equation In delta form, the scheme reads

$$-\frac{1}{\Delta x^2} \delta u_{i-1} + \frac{2}{\Delta x^2} \delta u_i = R_i.$$

Expanding about point (i, n)

$$-\frac{\Delta \tau}{\Delta x^2} \frac{\partial u_i^n}{\partial \tau} + \frac{\Delta \tau}{\Delta x} \frac{\partial^2 u_i^n}{\partial \tau \partial x} + O\left(\Delta \tau, \frac{\Delta \tau^2}{\Delta x^2}\right) + 2\frac{\Delta \tau}{\Delta x^2} \frac{\partial u_i^n}{\partial \tau} + O\left(\frac{\Delta \tau^2}{\Delta x^2}\right)$$

$$= \frac{\partial^2 u_i^n}{\partial x^2} + O(\Delta x^2) - f_i.$$

The leading time evolution equation corresponds to the heat equation again.

Consistency, Accuracy The scheme is consistent with equation (B.1) by choosing $\Delta \tau = \Delta x^2$, and the TE reads

$$\epsilon_i^n = O\left(\frac{\Delta \tau}{\Delta x}, \Delta \tau, \Delta x^2\right) = O(\Delta \tau^{\frac{1}{2}}, \Delta x^2).$$

The method is twice as fast as the previous one, since the time step is doubled. During the transient, the scheme is of order one-half in τ and order two in x. At steady-state, the scheme is second-order accurate. The solution is identical to the previous one.

B.1.3 Implicit Scheme

Making the Scheme Implicit All three values of u are moved to the new level. The computational molecule is shown in Fig. B.2.

Associated Matrix The system can be expressed as $T(p, q, r).u = s$. The matrix T is tridiagonal with coefficients and right-hand side given by:

$$p_i = -\frac{1}{\Delta x^2}, \quad q_i = \frac{2}{\Delta x^2}, \quad r_i = -\frac{1}{\Delta x^2}, \quad s_i = R_i.$$

The Thomas algorithm can be used to solve the system. The matrix T is diagonally dominant.

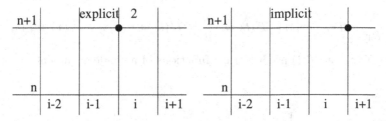

Fig. B.2. Sketch of computational molecule

Convergence The solution is obtained in one step, independent of the initial condition, since the system is linear and the solution method is exact.

B.2 Solution to Problem 2

B.2.1 Analytic Study

Equation Type Letting $v = \partial u/\partial t$, the second-order PDE is equivalent to the system of two first-order PDEs

$$\begin{cases} \dfrac{\partial u}{\partial t} = v \\ \dfrac{\partial v}{\partial x} = f(x) \end{cases}.$$

In matrix form, this is

$$\begin{bmatrix} 1 & 0 \\ 0 & 0 \end{bmatrix} \cdot \frac{\partial}{\partial t} \begin{bmatrix} u \\ v \end{bmatrix} + \begin{bmatrix} 0 & 0 \\ 0 & 1 \end{bmatrix} \cdot \frac{\partial}{\partial x} \begin{bmatrix} u \\ v \end{bmatrix} = \begin{bmatrix} v \\ f(x) \end{bmatrix}.$$

The characteristic matrix is

$$A = \begin{bmatrix} \dfrac{\partial \phi}{\partial t} & 0 \\ 0 & \dfrac{\partial \phi}{\partial x} \end{bmatrix},$$

and the characteristic form $Q = (\partial \phi / \partial t)(\partial \phi / \partial x)$. The characteristic lines have normal vectors $\nabla \phi = (0,1)$ and $(1,0)$. The system is totally hyperbolic.

Characteristic Lines The first family of characteristics is $\phi(x,t) = t = $ const, the second family is $\phi(x,t) = x = $ const. The information on the characteristics passing through point M travel in the direction of the arrows shown in Figure B.3.

General Solution The general solution to the homogeneous equation is obtained by two quadratures as

$$\frac{\partial v}{\partial x} = 0 \Rightarrow v(x,t) = T'(t) \Rightarrow \frac{\partial u}{\partial t} = T'(t) \Rightarrow u(x,t) = X(x) + T(t),$$

where $X(x)$ and $T(t)$ are arbitrary functions of a single argument.

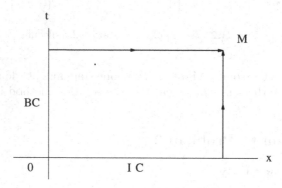

Fig. B.3. Characteristic lines and direction of propagation

Exact Solution Proceeding as before, for the non-homogeneous equation yields

$$\frac{\partial v}{\partial x} = f(x) \Rightarrow v(x,t) = \frac{\partial u}{\partial t} = \int_0^x f(\xi)d\xi + T'(t),$$

where the arbitrary constant of integration is included in T'. But at $x = 0$, $(\partial u / \partial t)(0,t) = h'(t) = T'(t)$, hence the second quadrature gives

$$u(x,t) = t \int_0^x f(\xi)d\xi + h(t) + X(x).$$

Now, applying the initial condition

$$u(x,0) = g(x) = h(0) + X(x),$$

which determines fully the solution as

$$u(x,t) = t \int_0^x f(\xi)d\xi + g(x) + h(t) - h(0).$$

Example If $u(x,t) = (1 + x + x^2)(1 + t)$ is the exact solution, then $f(x) = 1 + 2x$, $g(x) = 1 + x + x^2$, $h(t) = 1 + t$. Porting these in the above formula gives

$$u(x,t) = t \int_0^x (1 + 2\xi)d\xi + 1 + x + x^2 + 1 + t - 1 = t(x + x^2) + 1 + x + x^2 + t$$

$$= (1 + x + x^2)(1 + t).$$

B.2.2 Numerical Scheme

Type of Scheme This is not an implicit scheme since it can be marched in the x-direction with increasing values of the index i, the needed data being given at $x = 0$ to initiate the procedure.

Consistency, Accuracy The TE analysis is best carried out about point $(i - \frac{1}{2}, n + \frac{1}{2})$, to take advantage of the symmetries. Rewrite the scheme as

$$\epsilon_{i-\frac{1}{2}}^{n+\frac{1}{2}} = \frac{1}{\Delta t}\left(\frac{u_i^{n+1} - u_{i-1}^{n+1}}{\Delta x} - \frac{u_i^n - u_{i-1}^n}{\Delta x}\right) - f_{i-\frac{1}{2}}$$

$$= \frac{1}{\Delta t}\left(\frac{\partial u_{i-\frac{1}{2}}^{n+1}}{\partial x} - \frac{\partial u_{i-\frac{1}{2}}^n}{\partial x} + \frac{\Delta x^2}{24}\frac{\partial^3 u_{i-\frac{1}{2}}^{n+1}}{\partial x^3} - \frac{\Delta x^2}{24}\frac{\partial^3 u_{i-\frac{1}{2}}^n}{\partial x^3} + O(\Delta x^4)\right) - f_{i-\frac{1}{2}}$$

$$= \frac{\partial^2 u_{i-\frac{1}{2}}^{n+\frac{1}{2}}}{\partial t\partial x} - f_{i-\frac{1}{2}} + \frac{\Delta x^2}{24}\frac{\partial^4 u_{i-\frac{1}{2}}^{n+\frac{1}{2}}}{\partial t\partial x^3} + \frac{\Delta t^2}{24}\frac{\partial^4 u_{i-\frac{1}{2}}^{n+\frac{1}{2}}}{\partial t^3\partial x} + O(\Delta x^2\Delta t^2)$$

$$= O(\Delta t^2, \Delta x^2).$$

The scheme is consistent and second-order accurate in time and space.

Stability The FDE is made homogeneous by letting $f_{i-\frac{1}{2}} = 0$. The amplification factor is found to be $g = 1$. the scheme is unconditionally stable.

Example The solution in this example is an exact solution of the FDE. It is clear from the TE, where the Taylor expansions have been carried out far enough to indicate that the leading terms in the error are proportional to the fourth partial derivatives of u which are zero in this case. Another proof is to compute the TE according to the definition, by inserting the exact solution in the scheme. One finds indeed $\epsilon = 0$.

B.2.3 Numerical Application

Computational Molecule See Fig B.4. After multiplying by $\Delta t\Delta x$, the scheme, in update form, reads

$$u_i^{n+1} = u_{i-1}^{n+1} + u_i^n - u_{i-1}^n + 1.$$

In Fig. B.4, a local notation is introduced for the values at each point of the computational molecule. Let $e = \Delta t\Delta x f_{i-\frac{1}{2}}$, the scheme can be rewritten as

$$u_i^{n+1} = a = b + c - d + e.$$

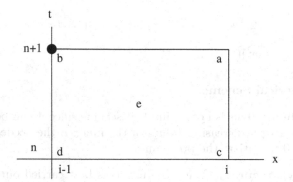

Fig. B.4. Computational molecule and local notation

Discrete Solution The 5×5 block of data is shown in Fig. B.5.

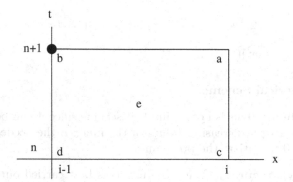

Fig. B.5. Computed block of data

Discussion The exact solution for this application is given by the result of 2.1.4 as $u(x,t) = (1 + x)(1 + t)$. The discrete solution corresponds to $u_i^n = (1 + x_i)(1 + t^n)$. This is in agreement with the truncation error which is zero also for this case.

B.3 Solution to Problem 3

B.3.1 Analytic Study

Equation Type The equivalent first-order system to the given second-order PDE reads:

$$\begin{cases} \dfrac{\partial \varphi}{\partial x} = u \\[2mm] \dfrac{\partial u}{\partial t} + c\dfrac{\partial u}{\partial x} = 0 \end{cases}.$$

Casting this system in matrix form yields

$$\begin{bmatrix} 0 & 0 \\ 0 & 1 \end{bmatrix} \cdot \frac{\partial}{\partial t}\begin{bmatrix} \varphi \\ u \end{bmatrix} + \begin{bmatrix} 1 & 0 \\ 0 & c \end{bmatrix} \cdot \frac{\partial}{\partial x}\begin{bmatrix} \varphi \\ u \end{bmatrix} = \begin{bmatrix} u \\ 0 \end{bmatrix}.$$

The characteristic matrix is

$$A = \begin{bmatrix} \dfrac{\partial \phi}{\partial x} & 0 \\[3mm] 0 & \dfrac{\partial \phi}{\partial t} + c\dfrac{\partial \phi}{\partial x} \end{bmatrix},$$

and its determinant is the characteristic form $Q = (\partial \phi / \partial x)((\partial \phi / \partial t) + c(\partial \phi / \partial x))$.

The characteristic polynomial $Q = 0$ has two real roots. There exists two distinct characteristic directions

$$\begin{cases} C^+ : \ \left(\dfrac{dx}{dt}\right)_{C^+} = c \\[3mm] C^\infty : \ \left(\dfrac{dx}{dt}\right)_{C^\infty} = \infty \end{cases}.$$

The characteristics are the straight lines

$$\begin{cases} C^+ : \ \phi(x,t) = x - ct = \text{const} \\[2mm] C^\infty : \ \phi(x,t) = t = \text{const} \end{cases}.$$

The equation and the system are totally hyperbolic.

General Solution The second first-order PDE is the linear convection equation. Its general solution is $u(x,t) = F'(x - ct)$.

Integrating once more gives $\varphi(x,t) = F(x - ct) + G(t)$, where F and G are arbitrary functions of a single argument.

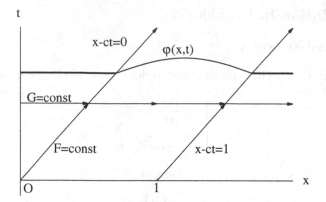

Fig. B.6. Subdomains and solution

Exact Solution The three regions of interest are bounded by the two characteristics $\xi = x - ct = 0$, $\xi = 1$. See Fig. B.6.

In region 1 we apply the boundary conditions

$$
\begin{cases}
\varphi(0,t) = F(-ct) + G(t) = 0 \\[2mm]
\dfrac{\partial \varphi}{\partial x}(0,t) = F'(-ct) + G'(t) = 0
\end{cases}
\forall t \geq 0.
$$

Clearly this implies that F and G are constant and without restriction can be chosen to be zero.

In region 1, $\varphi(x,t) = 0$, $x - ct \leq 0$, $t \geq 0$.

This results in F being zero on the characteristic $\xi = 0$ and G being zero everywhere.

In region 2 we use the initial condition

$$\varphi(x,0) = \sin \pi x, \ 0 \leq x \leq 1.$$

This determines the unknown function F in this region as $F(\xi) = \sin \pi \xi$, $0 \leq \xi \leq 1$. Thus in region 1 the solution is $\varphi(x,t) = \sin \pi(x - ct)$, $0 \leq x - ct \leq 1$. Note the continuity of φ at $\xi = 0$.

In region 3 the initial condition indicates that $F = 0$. Hence $\varphi(x,t) = 0$, $x - ct \geq 1$. φ is continuous at $\xi = 1$.

B.3.2 Numerical Study

Consistency, Accuracy This is not an implicit scheme. The solution is obtained in marching in the x-direction with increasing i.

The TE is defined as

$$
\epsilon_i^n = \frac{\varphi_i^{n+1} - \varphi_{i-1}^{n+1} - \varphi_i^n + \varphi_{i-1}^n}{\Delta t \Delta x} + c \frac{\varphi_i^n - 2\varphi_{i-1}^n + \varphi_{i-2}^n}{\Delta x^2}.
$$

Before expanding, we can rewrite the first term as

$$
\frac{\dfrac{\varphi_i^{n+1} - \varphi_{i-1}^{n+1}}{\Delta x} - \dfrac{\varphi_i^n - \varphi_{i-1}^n}{\Delta x}}{\Delta t},
$$

to recognize FD quotients in the numerator for which the Taylor expansions are known. The first quotient yields

$$
\frac{\varphi_i^{n+1} - \varphi_{i-1}^{n+1}}{\Delta x} = \frac{\partial \varphi_i^{n+1}}{\partial x} - \frac{\Delta x}{2}\frac{\partial^2 \varphi_i^{n+1}}{\partial x^2} + \frac{\Delta x^2}{3!}\frac{\partial^3 \varphi_i^{n+1}}{\partial x^3} + O(\Delta x^3).
$$

We now shift each term to (i, n), to get:

$$
= \frac{\partial \varphi_i^n}{\partial x} + \Delta t \frac{\partial^2 \varphi_i^n}{\partial t \partial x} + \frac{\Delta t^2}{2}\frac{\partial^3 \varphi_i^n}{\partial t^2 \partial x} + O(\Delta t^3)
$$

$$
- \frac{\Delta x}{2}\left(\frac{\partial^2 \varphi_i^n}{\partial x^2} + \Delta t \frac{\partial^3 \varphi_i^n}{\partial t \partial x^2} + O(\Delta t^2)\right)
$$

$$
+ \frac{\Delta x^2}{3!}\left(\frac{\partial^3 \varphi_i^n}{\partial x^3} + O(\Delta t)\right) + O(\Delta x^3).
$$

Now, the complete first term can be evaluated and the TE for the scheme reads

$$
\epsilon_i^n = \frac{\partial^2 \varphi_i^n}{\partial t \partial x} + \frac{3\Delta t}{2}\frac{\partial^3 \varphi_i^n}{\partial t^2 \partial x} - \frac{\Delta x}{2}\frac{\partial^3 \varphi_i^n}{\partial t \partial x^2} + O(\Delta t + \Delta x)^2
$$

$$
+ c\frac{\partial^2 \varphi_i^n}{\partial x^2} - c\Delta x\frac{\partial^3 \varphi_i^n}{\partial x^3} + O(\Delta x^2)
$$

$$
= O(\Delta t, \Delta x).
$$

Note that in order to get the remainder to be $O(\Delta t + \Delta x)^2$ the Taylor expansions had to be carried out to $O(\Delta t + \Delta x)^3$ in the intermediary calculations.

The scheme is consistent and first-order accurate in time and space.

Stability Let $\sigma = c(\Delta t/\Delta x) > 0$, $\varphi_i^n = g^n e^{ii\alpha}$. One finds

$$
g = \frac{(1 - \cos\alpha)(1 + 2\sigma\cos\alpha) + i\sin\alpha\,(1 - 2\sigma(1 - \cos\alpha))}{1 - \cos\alpha + i\sin\alpha}.
$$

The amplification factor is complex. We take its modulus:

$$
|g|^2 = \frac{(1 - \cos\alpha)^2(1 + 2\sigma\cos\alpha)^2 + \sin\alpha^2\,(1 - 2\sigma(1 - \cos\alpha))^2}{(1 - \cos\alpha)^2 + \sin\alpha^2}.
$$

After some algebra one finds

$$
|g|^2 = 1 - 2\sigma(1 - \sigma)(1 - \cos\alpha).
$$

$|g|^2 \leq 1$ is verified when $\sigma \leq 1$. This is a CFL condition for this explicit scheme. In terms of Δt this is

$$
\Delta t \leq \frac{\Delta x}{c}.
$$

B.3.3 Implicit Scheme

See the sketch of the computational molecule in Fig. B.7.

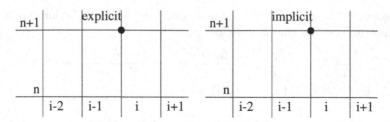

Fig. B.7. Sketch of computational molecule

By symmetry with the explicit scheme (exchanging n and $n+1$) the TE ϵ_i^{n+1} will have the same form as the previous TE. The scheme is consistent and first-order accurate in time and space.

The amplification factor is now

$$g_{\text{implicit}} = \frac{1 - \cos\alpha + i\sin\alpha}{(1 - \cos\alpha)(1 - 2\sigma\cos\alpha) + i\sin\alpha\,(1 + 2\sigma(1 - \cos\alpha))}$$

$$= \frac{1}{g_{\text{explicit}}(-\sigma)}.$$

Using the previous result, one finds

$$|g|^2_{\text{implicit}} = \frac{1}{1 + 2\sigma(1 + \sigma)(1 - \cos\alpha)} \leq 1, \ \forall\alpha.$$

The implicit scheme is unconditionally stable.

B.4 Solution to Problem 4

B.4.1 Explicit Scheme

Equation Type Let $v = \partial u/\partial x$. The second-order PDE can be replaced by the equivalent system of two first-order PDEs as

$$\begin{cases} \dfrac{\partial u}{\partial x} = v \\ \dfrac{\partial u}{\partial t} - \dfrac{\partial}{\partial x}\,(\nu(u)v) = 0 \end{cases}.$$

To study the system type it is necessary to use the quasi-linear form. Here it reads

$$\begin{cases} \dfrac{\partial u}{\partial x} = v \\ \dfrac{\partial u}{\partial t} - \nu(u)\dfrac{\partial v}{\partial x} = \nu'(x)v^2 \end{cases}.$$

Casting the system in matrix form yields

$$\begin{bmatrix} 0 & 0 \\ 1 & 0 \end{bmatrix} \cdot \frac{\partial}{\partial t} \begin{bmatrix} u \\ v \end{bmatrix} + \begin{bmatrix} 1 & 0 \\ 0 & -\nu(u) \end{bmatrix} \cdot \frac{\partial}{\partial x} \begin{bmatrix} u \\ v \end{bmatrix} = \begin{bmatrix} v \\ \nu'(u)v^2 \end{bmatrix}.$$

The characteristic matrix is

$$A = \begin{bmatrix} \dfrac{\partial \phi}{\partial x} & 0 \\ \dfrac{\partial \phi}{\partial t} & -\nu(u)\dfrac{\partial \phi}{\partial x} \end{bmatrix},$$

and the characteristic form $Q = -\nu(u)\left(\partial\phi/\partial x\right)^2$. The characteristic directions are given by $\partial\phi/\partial x = 0$, a double root. This is similar to the heat equation. the system is parabolic.

Characteristic Curves The characteristic lines have equation $\phi(x,t) = t = $ const. They are straight lines. See Fig. B.8.

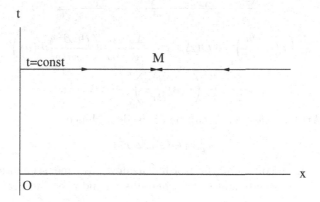

Fig. B.8. Characteristic lines of the PDE

Characteristic Speed The direction of propagation on the characteristic lines is in both directions. This is the reason for the double root. It allows to specify boundary conditions at both ends of the domain.

The speed of propagation is $(dx/dt)_{C\pm} = \pm\infty$.

Consistency, Accuracy We choose (i,n) to be the point for the study of the TE.

The term in the left-hand side expands as

$$\frac{u_i^{n+1} - u_i^n}{\Delta t} = \frac{\partial u_i^n}{\partial t} + \frac{\Delta t}{2} \frac{\partial^2 u_i^n}{\partial t^2} + O(\Delta t^2).$$

We need some intermediate results for averaged quantities which are more conveniently expanded about the mid-point, such as

$$\frac{u_i^n + u_{i+1}^n}{2} = u_{i+\frac{1}{2}}^n + \frac{\Delta x^2}{8} \frac{\partial^2 u_{i+\frac{1}{2}}^n}{\partial x^2} + O(\Delta x^4).$$

It follows that the value of ν, evaluated at the average point, is

$$\nu_{\overline{i+\frac{1}{2}}} = \nu \left(\frac{u_i^n + u_{i+1}^n}{2} \right) = \nu_{i+\frac{1}{2}} + \frac{\Delta x^2}{8} \frac{\partial^2 u_{i+\frac{1}{2}}^n}{\partial x^2} \nu'_{i+\frac{1}{2}} + O(\Delta x^4).$$

Finally we need the mid-point evaluation of the first derivatives such as

$$\frac{u_{i+1}^n - u_i^n}{\Delta x} = \frac{\partial u_{i+\frac{1}{2}}^n}{\partial x} + \frac{\Delta x^2}{24} \frac{\partial^3 u_{i+\frac{1}{2}}^n}{\partial x^3} + O(\Delta x^4).$$

The Taylor expansion of the term in the right-hand side can be obtained:

$$\frac{\nu_{\overline{i+\frac{1}{2}}} \dfrac{u_{i+1}^n - u_i^n}{\Delta x} - \nu_{\overline{i-\frac{1}{2}}} \dfrac{u_i^n - u_{i-1}^n}{\Delta x}}{\Delta x} =$$

$$\frac{\partial}{\partial x} \left(\nu(u) \frac{\partial u}{\partial x} \right)_i^n + O(\Delta x^2) + \frac{\Delta x^2}{8} \frac{\partial}{\partial x} \left(\frac{\partial u}{\partial x} \frac{\partial^2 u}{\partial x^2} \nu'(u) \right)_i^n$$

$$+ \frac{\Delta x^2}{24} \frac{\partial}{\partial x} \left(\nu(u) \frac{\partial^3 u}{\partial x^3} \right)_i^n + O(\Delta x^4).$$

To $O(\Delta x^2)$ the last terms can be discarded. Hence

$$\epsilon_i^n = O(\Delta t, \Delta x^2).$$

The scheme is consistent, first-order accurate in time, second-order accurate in space. At steady-state, the scheme is second-order accurate in x.

Stability Let $\nu = \text{const}$, $r = \nu(\Delta t/\Delta x^2)$, $u_i^n = g^n e^{ii\alpha}$. The amplification factor is found to be

$$g = 1 - 2r(1 - \cos\alpha).$$

g is real. For stability one needs $-1 \leq g \leq 1$. The left inequality leads to the stability condition $r \leq \frac{1}{2}$. In terms of Δt, this is

$$\Delta t \leq \frac{\Delta x^2}{2\nu}.$$

This scheme corresponds to Jacobi's method.

B.4.2 Semi-Implicit Scheme

Equation Type The quasi-linear first-order system is now

$$
\begin{cases}
\dfrac{\partial u}{\partial x} = v \\[2mm]
\dfrac{\partial v}{\partial t} - \nu(u)\dfrac{\partial v}{\partial x} = \nu'(u)v^2
\end{cases}.
$$

The matrix form of the system follows

$$
\begin{bmatrix} 0 & 0 \\ 0 & 1 \end{bmatrix} \cdot \frac{\partial}{\partial t} \begin{bmatrix} u \\ v \end{bmatrix} + \begin{bmatrix} 1 & 0 \\ 0 & -\nu(u) \end{bmatrix} \cdot \frac{\partial}{\partial x} \begin{bmatrix} u \\ v \end{bmatrix} = \begin{bmatrix} v \\ \nu'(u)v^2 \end{bmatrix}.
$$

The characteristic matrix is

$$
A = \begin{bmatrix} \dfrac{\partial \phi}{\partial x} & 0 \\[3mm] 0 & \dfrac{\partial \phi}{\partial t} - \nu(u)\dfrac{\partial \phi}{\partial x} \end{bmatrix}.
$$

Taking the determinant of A one gets the characteristic form $Q = (\partial\phi/\partial x)\left((\partial\phi/\partial t) - \nu(u)(\partial\phi/\partial x)\right)$. There are two distinct roots to the characteristic form. The system is totally hyperbolic.

Characteristic Curves The characteristic directions are

$$
\left(\frac{dx}{dt}\right)_C = -\frac{\dfrac{\partial \phi}{\partial t}}{\dfrac{\partial \phi}{\partial x}}.
$$

The two characteristic directions are

$$
\begin{cases}
C^\infty : \left(\dfrac{dx}{dt}\right)_{C^\infty} = \infty \\[3mm]
C^- : \left(\dfrac{dx}{dt}\right)_{C^-} = -\nu(u)
\end{cases}.
$$

The first family of characteristics are the straight lines $C^\infty : \phi(x,t) = t = $ const.

The second family, the C^--lines, are curved lines that depend on the solution. See Fig. B.9.

Characteristic Speed The characteristic speed of perturbations for the C^∞-lines is infinite.

For the C^--lines, the speed is $-\nu(u)$.

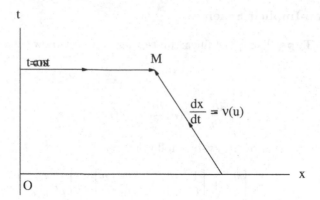

Fig. B.9. Characteristic directions for the PDE

Consistency, Accuracy We elect to expand the TE about the point (i, n), as before. The right hand-side is unchanged. The Taylor expansion of the left hand-side reads:

$$\frac{\dfrac{u_i^{n+1} - u_{i-1}^{n+1}}{\Delta x} - \dfrac{u_i^n - u_{i-1}^n}{\Delta x}}{\Delta t} =$$

$$\frac{\left(\dfrac{\partial u}{\partial x} - \dfrac{\Delta x}{2}\dfrac{\partial^2 u}{\partial x^2} + \dfrac{\Delta x^2}{3!}\dfrac{\partial^3 u}{\partial x^3} + O(\Delta x^3)\right)_i^{n+1} - \left(\dfrac{\partial u}{\partial x} - \dfrac{\Delta x}{2}\dfrac{\partial^2 u}{\partial x^2} + \dfrac{\Delta x^2}{3!}\dfrac{\partial^3 u}{\partial x^3} + O(\Delta x^3)\right)_i^n}{\Delta t}$$

$$= \frac{\partial^2 u_i^n}{\partial t \partial x} + \frac{\Delta t}{2}\frac{\partial^3 u_i^n}{\partial t^2 \partial x} - \frac{\Delta x}{2}\frac{\partial^3 u_i^n}{\partial t \partial x^2} + O(\Delta t + \Delta x)^2.$$

The result for the TE is

$$\epsilon_i^n = O(\Delta t, \Delta x, \Delta x^2).$$

Note that the first-order error in Δx comes from a time derivative term. At steady-state, the scheme is second-order accurate in x.

This is not an implicit scheme. The solution can be found by marching in the x-direction. This scheme corresponds to Gauss-Seidel's method.

Stability Let $\sigma = \nu(\Delta t/\Delta x)$. Using the linearized version of the right-hand side, $\nu = $ const, and using the same definitions of the complex modes, one gets:

$$g = \frac{(1 - \cos\alpha)(1 - 2\sigma) + i\sin\alpha}{1 - \cos\alpha + i\sin\alpha} = \frac{a + ib}{c + ib}.$$

The amplification factor is a complex number. Its modulus will be less than one if $|a| \leq |c|$. This is satisfied when $\sigma \leq 1$. In terms of Δt the condition is

$$\Delta t \leq \frac{\Delta x}{\nu}.$$

This is one order of magnitude faster than the explicit scheme.

B.4.3 Implicit Scheme

We use a fixed-point linearization. The coefficients $\nu_{i\pm\frac{1}{2}}$ are evaluated at the "old" level. The derivatives $\partial u/\partial x$ are evaluated at the "new" level as

$$\frac{\nu_{i+\frac{1}{2}}\dfrac{u_{i+1}^{n+1} - u_i^{n+1}}{\Delta x} - \nu_{i-\frac{1}{2}}\dfrac{u_i^{n+1} - u_{i-1}^{n+1}}{\Delta x}}{\Delta x} = 0.$$

The computational molecule is sketched in Fig. B.10.

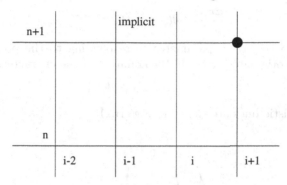

Fig. B.10. Sketch of computational molecule

The matrix of the fixed point iterative method is tridiagonal and can be written as $Au^{n+1} = b$, where $A = T(p, q, r)$

$$\begin{cases} p = \nu_{i-\frac{1}{2}} \\ q = -\nu_{i-\frac{1}{2}} - \nu_{i+\frac{1}{2}} \\ r = \nu_{i+\frac{1}{2}} \end{cases} , \quad i = 2, ..., ix - 1.$$

The system can be solved easily using the Thomas algorithm. The matrix is diagonally dominant.

It is necessary to iterate since the equation is nonlinear.

B.5 Solution to Problem 5

B.5.1 Analytic Study

Equation Type To study the equation type, we expand the PDE in non-conservative form as

$$\frac{\partial u}{\partial t} - x\frac{\partial u}{\partial x} = u.$$

The characteristic form reads

$$Q = \frac{\partial \phi}{\partial t} - x\frac{\partial \phi}{\partial x}.$$

There is always a real root. The equation is hyperbolic.

The PDE is linear. If u_1 and u_2 are solutions, λu_1 and $u_1 + u_2$ are also solutions.

Characteristic Lines The slope of the characteristic line is given by

$$\left(\frac{dx}{dt}\right)_C = -\frac{\partial \phi}{\partial t} \bigg/ \frac{\partial \phi}{\partial x} = -x.$$

Integrating gives

$$\frac{dx}{x} = -dt \Rightarrow x_C(t) = Ce^{-t}.$$

If we introduce x_0, a parameter corresponding to the point where the characteristic originates at $t = 0$, the equation of the characteristic becomes

$$x_C(t) = x_0 e^{-t}.$$

Characteristic lines are shown in Fig. B.11.

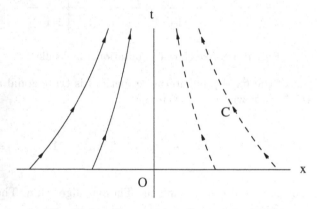

Fig. B.11. Characteristic lines of the PDE

Jump Condition The jump condition is

$$\langle u \rangle \, n_t - \langle xu \rangle \, n_x = 0$$

Since x is a continuous variable at the shock S, i.e. $\langle x \rangle = 0$, the jump condition reduces to $\langle u \rangle \, n_t - x \langle u \rangle \, n_x = 0$. For a non-trivial jump, i.e. $\langle u \rangle \neq 0$, this reduces further to $n_t - x n_x = 0$. The slope of the jump line is

$$\left(\frac{dx}{dt}\right)_S = -\frac{n_t}{n_x} = -x = \left(\frac{dx}{dt}\right)_C.$$

The shock line coincides with a characteristic. This is a general result for linear equations.

General Solution Let $u_C\left(x_C(t), t\right)$ be the value of $u(x,t)$ on the characteristic C. Taking the derivative in time gives

$$\frac{du_C}{dt} = \frac{\partial u}{\partial t} + \left(\frac{dx}{dt}\right)_C \frac{\partial u}{\partial x} = \frac{\partial u}{\partial t} - x\frac{\partial u}{\partial x} = u_C.$$

Integrating this ODE gives the general solution to the PDE as

$$u(x,t) = C(\xi)e^t, \ \xi = xe^t = x_0.$$

$C(\xi)$ is an arbitrary function of the single argument ξ. $\xi = \text{const}$ on characteristics.

Exact Solution given the initial condition, we find $C(x) = f(x)$. The exact solution is

$$u(x,t) = e^t f(xe^t).$$

Checking the result we find

$$\frac{\partial u}{\partial t} = xe^{2t} f' + e^t f, \ \frac{\partial u}{\partial x} = e^{2t} f' \Rightarrow \frac{\partial u}{\partial t} - x\frac{\partial u}{\partial x} = e^t f = u.$$

B.5.2 Numerical Study

B.5.3 Discrete Conservation Form

Consider a natural conservative discretization such as

$$\frac{u_i^{n+1} - u_i^n}{\Delta t} - \frac{x_i u_i^n - x_{i-1} u_{i-1}^n}{\Delta x} = 0.$$

Since $x_i = x_{i-\frac{1}{2}} + (\Delta x/2)$, and $x_{i-1} = x_{i-\frac{1}{2}} - (\Delta x/2)$, one can write

$$\frac{u_i^{n+1} - u_i^n}{\Delta t} - x_{i-\frac{1}{2}} \frac{u_i^n - u_{i-1}^n}{\Delta x} - \frac{u_{i-1}^n + u_i^n}{2} = 0.$$

This is scheme i). it is possible to proceed in a similar manner for schemes iv). Scheme ii) will be a combination of schemes i) and iv). Scheme iii) is not a conservative scheme. It cannot be written as a discrete conservation form.

B.5.4 Consistency

The truncation error yields

$$\epsilon_{i-\frac{1}{2}}^n = \left\{\frac{\partial u}{\partial t} + \frac{\Delta t}{2}\frac{\partial^2 u}{\partial t^2} + \frac{\Delta t^2}{6}\frac{\partial^3 u}{\partial t^3} + O(\Delta t^3)\right\}_i^n$$

$$-\left\{x\left[\frac{\partial u}{\partial x} + \frac{\Delta x^2}{24}\frac{\partial^3 u}{\partial x^3} + O(\Delta x^4)\right]\right\}_{i-\frac{1}{2}}^n - \left\{u + \frac{\Delta x^2}{8}\frac{\partial^2 u}{\partial x^2} + O(\Delta x^4)\right\}_{i-\frac{1}{2}}^n$$

There remains to re-expand the first bracket about point $i - \frac{1}{2}$. Using the Taylor formula

$$\{\}_i^n = \{\}_{i-\frac{1}{2}}^n + \frac{\Delta x}{2}\frac{\partial}{\partial x}\{\}_{i-\frac{1}{2}}^n + \frac{\Delta x^2}{8}\frac{\partial^2}{\partial x^2}\{\}_{i-\frac{1}{2}}^n + O(\Delta x^3)$$

results in

$$\epsilon_{i-\frac{1}{2}}^n = \left\{\frac{\partial u}{\partial t} - x\frac{\partial u}{\partial x} - u\right\}_{i-\frac{1}{2}}^n + \frac{\Delta t}{2}\frac{\partial^2 u}{\partial t^2}_{i-\frac{1}{2}}^n + \frac{\Delta x}{2}\frac{\partial^2 u}{\partial t\partial x}_{i-\frac{1}{2}}^n + \frac{\Delta t^2}{6}\frac{\partial^3 u}{\partial t^3}_{i-\frac{1}{2}}^n$$

$$+\frac{\Delta t\Delta x}{4}\frac{\partial^3 u}{\partial t^2\partial x}_{i-\frac{1}{2}}^n + \Delta x^2\left\{\frac{1}{8}\frac{\partial^3 u}{\partial t\partial x^2} - \frac{x}{24}\frac{\partial^3 u}{\partial x^3} - \frac{1}{8}\frac{\partial^2 u}{\partial x^2}\right\}_{i-\frac{1}{2}}^n$$

$$+O(\Delta t^3, \Delta t^2\Delta x, \Delta t\Delta x^2, \Delta x^3) = O(\Delta t, \Delta x, \Delta x^2)$$

The terms in the first bracket vanish since u is the exact solution. The scheme is consistent, first-order accurate in time and space for the transient. At steady-state the scheme is second-order accurate in space.

B.5.5 Stability

The linear stability method of Von-Neumann is applied to the homogeneous equation with $u_i^n = g^n e^{ii\alpha}$ to yield

$$g - 1 + \sigma(1 - e^{-i\alpha}) = 0$$

where $\sigma = -x_{i-\frac{1}{2}}(\Delta t/\Delta x) \geq 0$. Solving for g results in

$$g = 1 - \sigma(1 - \cos\alpha) - \underline{i}\sigma\sin\alpha$$

The modulus of the amplification factor g is

$$|g|^2 = [1 - \sigma(1 - \cos\alpha)]^2 + [\sigma\sin\alpha]^2 = 1 - 2(1 - \cos\alpha)\sigma(1 - \sigma)$$

The stability condition reduces to $\sigma \leq 1$, i.e. $\Delta t \leq -(\Delta x/x_{i-\frac{1}{2}})$.
This condition may not be met for large values of $x_{i-\frac{1}{2}}$ or would require a prohibitively fine mesh.

B.5.6 Implicit scheme

The following implicit scheme is proposed

$$\frac{u_i^{n+1} - u_i^n}{\Delta t} - x_{i-\frac{1}{2}}\frac{u_i^{n+1} - u_{i-1}^{n+1}}{\Delta x} - \frac{u_{i-1}^n + u_i^n}{2} = 0$$

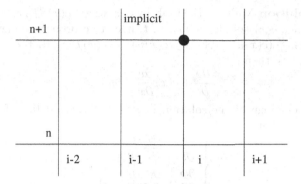

Fig. B.12. Sketch of computational molecule

This is sketched in Fig. B.12. Note that the source term has not been included in the molecules as it does not participate in the equation type nor stability.

The stability study of the homogeneous equation yields

$$g = \frac{1}{1 + \sigma(1 - \cos\alpha) + \underline{i}\sigma\sin\alpha}$$

indicating unconditional stability since the complex number in the denominator is always larger than one in modulus. The associated matrix is bi-diagonal (for $i < 0$) and reads

$$\begin{bmatrix} \cdot & O & O \\ \dfrac{x_{i-\frac{1}{2}}}{\Delta x}\dfrac{1}{\Delta t} - \dfrac{x_{i-\frac{1}{2}}}{\Delta x} & O & \\ O & \cdot & \cdot \end{bmatrix} \begin{bmatrix} \cdot \\ \delta u_i \\ \cdot \end{bmatrix} = \begin{bmatrix} \cdot \\ x_{i-\frac{1}{2}}\dfrac{u_i^n - u_{i-1}^n}{\Delta x} + \dfrac{u_{i-1}^n + u_i^n}{2} \\ \cdot \end{bmatrix}$$

This system can be solved with the Thomas algorithm.

B.6 Solution to Problem 6

B.6.1 Analytic Study

Equation Type In non-conservative form the PDE reads $(\partial u/\partial t) + c(x)(\partial u/\partial x) = f(x) - c'(x)u$. This is a linear convection equation with a source term. It is hyperbolic.

The characteristic direction is given by $(dx/dt)_C = c(x)$.

General Solution Multiply the PDE by $c(x)$ to get $c(x)\frac{\partial u}{\partial t}+c(x)\frac{\partial}{\partial x}(c(x)u)=c(x)f(x)$. Since $c(x)$ is independent of t, it can be brought under the partial derivative as $\frac{\partial}{\partial t}(c(x)u)+c(x)\frac{\partial}{\partial x}(c(x)u)=c(x)f(x)$. Let $v=c(x)u$ and $g(x)=c(x)f(x)$, then

$$\frac{\partial v}{\partial t}+c(x)\frac{\partial v}{\partial x}=g(x). \tag{B.2}$$

An implicit form of the solution is $\Phi(x,t,v(x,t))=0$. It follows, using the chain rule, that

$$\begin{cases} \dfrac{\partial \Phi}{\partial t}+\dfrac{\partial \Phi}{\partial v}\dfrac{\partial v}{\partial t}=0 \\[2mm] \dfrac{\partial \Phi}{\partial x}+\dfrac{\partial \Phi}{\partial v}\dfrac{\partial v}{\partial x}=0 \end{cases}.$$

Multiplying (B.2) by $\partial\Phi/\partial v$, and eliminating $\partial v/\partial t$ and $\partial v/\partial x$, using the two relations above, yields

$$\frac{\partial \Phi}{\partial t}+c(x)\frac{\partial \Phi}{\partial x}+g(x)\frac{\partial \Phi}{\partial v}=0.$$

Integrating $dt=dx/c(x)=dv/g(x)$ gives the solution $v(x,t)=\int^x(g(\xi)/c(\xi))d\xi+F\left(t-\int^x(1/c(\xi))d\xi\right)$, where F is an arbitrary function of a single argument. To verify that this is the general solution, one takes partial derivatives of v w.r. to t and x as

$$\frac{\partial v}{\partial t}=F',\ \frac{\partial v}{\partial x}=\frac{g(x)}{c(x)}-\frac{1}{c(x)}F'\Rightarrow \frac{\partial v}{\partial t}+c(x)\frac{\partial v}{\partial x}=F'+g(x)-F'=g(x).$$

Application Let $c(x)=2x-1$, $f(x)=4c(x)$. The characteristic lines are obtained by integration of $dx/dt=2x-1$. This results in $\phi(x,t)=t-\frac{1}{2}\ln|2x-1|=$ const. This is shown in Fig. B.13.

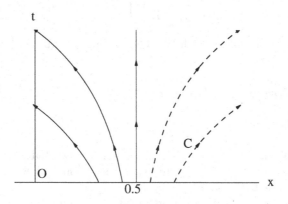

Fig. B.13. Characteristic lines of the PDE

B.6.2 Numerical Scheme

Consistency, Accuracy See Fig. B.14. At steady-state the schemes reduce to the space derivatives and the source terms.

i) $c_{i-\frac{1}{2}} > 0$ and $c_{i+\frac{1}{2}} > 0$

$$c_{i-\frac{1}{2}} \frac{u_i - u_{i-1}}{\Delta x} = 4c_{i-\frac{1}{2}} - 2u_{\overline{i-\frac{1}{2}}}.$$

We choose to expand about $i - \frac{1}{2}$. Some auxiliary Taylor expansions are needed:

$$\begin{cases} \dfrac{u_i - u_{i-1}}{\Delta x} = \dfrac{\partial u_{i-\frac{1}{2}}}{\partial x} + \dfrac{\Delta x^2}{24} \dfrac{\partial^3 u_{i-\frac{1}{2}}}{\partial x^3} + O(\Delta x^4) \\[2ex] u_{\overline{i-\frac{1}{2}}} = \dfrac{u_{i-1} + u_i}{2} = u_{i-\frac{1}{2}} + \dfrac{\Delta x^2}{8} \dfrac{\partial^2 u_{i-\frac{1}{2}}}{\partial x^2} + O(\Delta x^4) \end{cases}$$

One finds the TE to be

$$\epsilon_{i-\frac{1}{2}} = c_{i-\frac{1}{2}} \frac{\partial u_{i-\frac{1}{2}}}{\partial x} - 4c_{i-\frac{1}{2}} + 2u_{i-\frac{1}{2}} + \frac{\Delta x^2}{24} c_{i-\frac{1}{2}} \frac{\partial^3 u_{i-\frac{1}{2}}}{\partial x^3} + \frac{\Delta x^2}{4} \frac{\partial^2 u_{i-\frac{1}{2}}}{\partial x^2} + O(\Delta x^4)$$

$$= O(\Delta x^2).$$

The scheme is consistent and second-order accurate at steady-state.

It is clear that the same result holds for scheme iv), which is a mirror image of scheme i). The result carries over to scheme ii), which is the sum of scheme i) and iv), and therefore is consistent and second-order accurate at steady-state. Note, however, that scheme ii) is inconsistent in the transient, but the error vanishes with x.

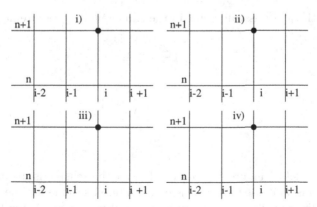

Fig. B.14. Computational molecules for the explicit scheme

Finally, scheme iii) is consistent and second-order accurate at steady-state, as is obvious due to the fact that all the schemes are centered about point i.

The steady-state solution is an exact solution of the FDE's. Note that $u_{ss\,\overline{i\pm\frac{1}{2}}} = u_{ss\,i\pm\frac{1}{2}} = c_{i\pm\frac{1}{2}}$. This is verified by computing the TE:

i)

$$\epsilon = c_{i-\frac{1}{2}}\frac{(2x_i - 1) - (2x_{i-1} - 1)}{\Delta x} - 2c_{i-\frac{1}{2}} = c_{i-\frac{1}{2}}\left(2\frac{x_i - x_{i-1}}{\Delta x} - 2\right) = 0.$$

Using a symmetry argument, it follows that scheme iv) computes the exact solution, which in turn implies that ii) is identically satisfied as well.

Finally, scheme iii) reads

iii)

$$\epsilon = c_i\frac{(2x_{i+1} - 1) - (2x_{i-1} - 1)}{2\Delta x} - 2c_i = c_i\left(\frac{x_{i+1} - x_{i-1}}{\Delta x} - 2\right) = 0.$$

Stability The stability analysis is carried out using Von Neumann analysis. The PDE is linear, but it is made homogeneous by canceling the source term. In cases i), ii) and iv) the equations are identical to the linearized version of Burgers' equation (inviscid):

i)

$$\frac{u_i^{n+1} - u_i^n}{\Delta t} + c_{i-\frac{1}{2}}\frac{u_i^n - u_{i-1}^n}{\Delta x} = 0,$$

ii)

$$\frac{u_i^{n+1} - u_i^n}{\Delta t} + c_{i-\frac{1}{2}}\frac{u_i^n - u_{i-1}^n}{\Delta x} + c_{i+\frac{1}{2}}\frac{u_{i+1}^n - u_i^n}{\Delta x} = 0,$$

iv)

$$\frac{u_i^{n+1} - u_i^n}{\Delta t} + c_{i+\frac{1}{2}}\frac{u_{i+1}^n - u_i^n}{\Delta x} = 0.$$

Hence, the analysis will result in the same stability requirements:

i)

$$c_{i-\frac{1}{2}}\frac{\Delta t}{\Delta x} \leq 1,$$

ii)

$$(c_{i-\frac{1}{2}} - c_{i+\frac{1}{2}})\frac{\Delta t}{\Delta x} \leq 1,$$

iv)

$$-c_{i+\frac{1}{2}}\frac{\Delta t}{\Delta x} \leq 1.$$

Case iii) yields the unexpected result that the scheme is unconditionally unstable, as is well known when using a centered scheme for the convection equation. However, the scheme has been found to give good result. The only explanation found is that the linear source term, $-2u_i^n$, contributes to the stability, when included in the Von Neumann analysis.

B.6.3 Implicit Scheme

An implicit scheme can be devised by moving the space derivatives from level n to level $n+1$. The scheme now reads:

  i) $c_{i-\frac{1}{2}} > 0$ and $c_{i+\frac{1}{2}} > 0$

$$\frac{u_i^{n+1} - u_i^n}{\Delta t} + c_{i-\frac{1}{2}} \frac{u_i^{n+1} - u_{i-1}^{n+1}}{\Delta x} = 4c_{i-\frac{1}{2}} - 2u_{i-\frac{1}{2}}^n,$$

  ii) $c_{i-\frac{1}{2}} > 0$ and $c_{i+\frac{1}{2}} < 0$

$$\frac{u_i^{n+1} - u_i^n}{\Delta t} + c_{i-\frac{1}{2}} \frac{u_i^{n+1} - u_{i-1}^{n+1}}{\Delta x} + c_{i+\frac{1}{2}} \frac{u_{i+1}^{n+1} - u_i^{n+1}}{\Delta x}$$

$$= 4\left(c_{i-\frac{1}{2}} + c_{i+\frac{1}{2}}\right) - 2\left(u_{i-\frac{1}{2}}^n + u_{i+\frac{1}{2}}^n\right),$$

  iii) $c_{i-\frac{1}{2}} < 0$ and $c_{i+\frac{1}{2}} > 0$

$$\frac{u_i^{n+1} - u_i^n}{\Delta t} + c_i \frac{u_{i+1}^{n+1} - u_{i-1}^{n+1}}{2\Delta x} = 4c_i^n - 2u_i^n,$$

  iv) $c_{i-\frac{1}{2}} < 0$ and $c_{i+\frac{1}{2}} < 0$

$$\frac{u_i^{n+1} - u_i^n}{\Delta t} + c_{i+\frac{1}{2}} \frac{u_{i+1}^{n+1} - u_i^{n+1}}{\Delta x} = 4c_{i+\frac{1}{2}} - 2u_{i+\frac{1}{2}}^n.$$

The computational molecules are shown in Fig. B.15.

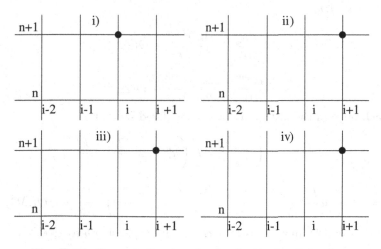

Fig. B.15. Computational molecules for the implicit scheme

The system of algebraic equations takes the form $A(p, q, r).u = s$, where A is a tridiagonal matrix. In the case $ix = 4$, the system reads

$$
\begin{bmatrix}
\dfrac{1}{\Delta t} - \dfrac{c_{\frac{3}{2}}}{\Delta x} & \dfrac{c_{\frac{3}{2}}}{\Delta x} & 0 & 0 \\[2ex]
-\dfrac{c_2}{\Delta x} & \dfrac{1}{\Delta t} & \dfrac{c_2}{\Delta x} & 0 \\[2ex]
0 & -\dfrac{c_3}{\Delta x} & \dfrac{1}{\Delta t} & \dfrac{c_3}{\Delta x} \\[2ex]
0 & 0 & -\dfrac{c_{\frac{7}{2}}}{\Delta x} & \dfrac{1}{\Delta t} + \dfrac{c_{\frac{7}{2}}}{\Delta x}
\end{bmatrix}
\cdot
\begin{bmatrix}
\delta u_1 \\[2ex] \delta u_2 \\[2ex] \delta u_3 \\[2ex] \delta u_4
\end{bmatrix}
=
\begin{bmatrix}
-c_{\frac{3}{2}} \dfrac{u_2^n - u_1^n}{\Delta x} + 4c_{\frac{3}{2}} - 2u_{\frac{3}{2}}^n \\[2ex]
-c_2 \dfrac{u_3^n - u_1^n}{2\Delta x} + 4c_2 - 2u_2^n \\[2ex]
-c_3 \dfrac{u_4^n - u_2^n}{2\Delta x} + 4c_3 - 2u_3^n \\[2ex]
-c_{\frac{7}{2}} \dfrac{u_4^n - u_3^n}{\Delta x} + 4c_{\frac{7}{2}} - 2u_{\frac{7}{2}}^n
\end{bmatrix}.
$$

The condition for the matrix to be diagonally dominant is $|c_{2,3}| (\Delta t/\Delta x) \leq 1$. this condition comes from the sonic points. It is not very restrictive since as the mesh is refined the coefficients c_i for the sonic points tend to $c\left(\frac{1}{2}\right) = 0$.

The Thomas algorithm can be used to solve the linear system.

B.7 Solution to Problem 7

B.7.1 The MacCormack Scheme

Consistency, Accuracy The MacCormack scheme can be written in PDE form as

$$
\frac{u_i^{n+1} - u_i^n}{\Delta t} + \frac{1}{2}\left(\frac{F_{i+1}^n - F_i^n}{\Delta x} + \frac{\overline{F_i^{n+1}} - \overline{F_{i-1}^{n+1}}}{\Delta x}\right) = 0.
$$

Let a,b and c be the three FD quotients. Each is expanded successively about point (i, n).

$$
a = \frac{\partial u_i^n}{\partial t} + \frac{\Delta t}{2}\frac{\partial^2 u_i^n}{\partial t^2} + O(\Delta t^2),
$$

$$
b = \frac{\partial F_i^n}{\partial x} + \frac{\Delta x}{2}\frac{\partial^2 F_i^n}{\partial x^2} + O(\Delta x^2).
$$

Each term in the last quotient is considered separately

$$
\overline{F_i^{n+1}} = F\left(\overline{u_i^{n+1}}\right) = F\left(u_i^n - \Delta t \frac{F_{i+1}^n - F_i^n}{\Delta x}\right)
$$

$$
= F_i^n - \Delta t \frac{F_{i+1}^n - F_i^n}{\Delta x}\frac{dF_i^n}{du} + \frac{\Delta t^2}{2}\left(\frac{F_{i+1}^n - F_i^n}{\Delta x}\right)^2\frac{d^2 F_i^n}{du^2} + O(\Delta t^3).
$$

Now, each FD quotient is replaced by its expansion b to give

$$F_i^{\overline{n+1}} = F_i^n - \Delta t \frac{\partial F_i^n}{\partial x}\frac{dF_i^n}{du} - \frac{\Delta t \Delta x}{2}\frac{\partial^2 F_i^n}{\partial x^2}\frac{dF_i^n}{du}$$

$$+ \frac{\Delta t^2}{2}\left(\frac{\partial F_i^n}{\partial x}\right)^2 \frac{d^2 F_i^n}{du^2} + O(\Delta t + \Delta x)^3.$$

The term $F_{i-1}^{\overline{n+1}}$ is obtained by shifting the Taylor expansion of $F_i^{\overline{n+1}}$ as

$$F_{i-1}^{\overline{n+1}} = F_i^n - \Delta x \frac{\partial F_i^n}{\partial x} + \frac{\Delta x^2}{2}\frac{\partial^2 F_i^n}{\partial x^2} + O(\Delta x^3)$$

$$- \Delta t \left(\frac{\partial F_i^n}{\partial x}\frac{dF_i^n}{du} - \Delta x \frac{\partial}{\partial x}\left(\frac{\partial F_i^n}{\partial x}\frac{dF_i^n}{du}\right) + O(\Delta x^2)\right)$$

$$- \frac{\Delta t \Delta x}{2}\left(\frac{\partial^2 F_i^n}{\partial x^2}\frac{dF_i^n}{du} + O(\Delta x)\right)$$

$$+ \frac{\Delta t^2}{2}\left(\frac{\partial F_i^n}{\partial x}\right)^2 \frac{d^2 F_i^n}{du^2} + O(\Delta t^2 \Delta x, \Delta t^3).$$

The FD quotient now reads

$$c = \frac{\partial F_i^n}{\partial x} - \frac{\Delta t}{2}\frac{\partial^2 F_i^n}{\partial x^2}\frac{dF_i^n}{du} - \frac{\Delta x}{2}\frac{\partial^2 F_i^n}{\partial x^2} - \Delta t \frac{\partial}{\partial x}\left(\frac{\partial F_i^n}{\partial x}\frac{dF_i^n}{du}\right) + O(\Delta t + \Delta x)^2.$$

The TE can now be derived:

$$\epsilon_i^n = \frac{\partial u_i^n}{\partial t} + \frac{\Delta t}{2}\frac{\partial^2 u_i^n}{\partial t^2} + O(\Delta t^2) + \frac{1}{2}\times$$

$$\left[\frac{\partial F_i^n}{\partial x} + \frac{\Delta x}{2}\frac{\partial^2 F_i^n}{\partial x^2} + \frac{\partial F_i^n}{\partial x} - \frac{\Delta t}{2}\frac{\partial^2 F_i^n}{\partial x^2}\frac{dF_i^n}{du}\right.$$

$$\left. - \frac{\Delta x}{2}\frac{\partial^2 F_i^n}{\partial x^2} - \Delta t \frac{\partial}{\partial x}\left(\frac{\partial F_i^n}{\partial x}\frac{dF_i^n}{du}\right)\right]$$

$$+ O(\Delta t + \Delta x)^2.$$

After some cancellations there remains

$$\epsilon_i^n = \frac{\Delta t}{2}\left(\frac{\partial^2 u_i^n}{\partial t^2} - \frac{\partial}{\partial x}\left(\frac{\partial F_i^n}{\partial x}\frac{dF_i^n}{du}\right)\right) + O(\Delta t + \Delta x)^2.$$

If we use the governing equation to evaluate $(\partial^2 u_i^n / \partial t^2) =$, then one finds

$$\frac{\partial^2 u_i^n}{\partial t^2} = \frac{\partial}{\partial t}\left(-\frac{\partial F_i^n}{\partial x}\right) = -\frac{\partial}{\partial x}\left(\frac{\partial F_i^n}{\partial t}\right)$$

$$= -\frac{\partial}{\partial x}\left(\frac{dF_i^n}{du}\frac{\partial u_i^n}{\partial t}\right) = \frac{\partial}{\partial x}\left(\frac{\partial F_i^n}{\partial x}\frac{dF_i^n}{du}\right).$$

Hence, the term in bracket in the TE is zero. The scheme is second-order accurate in t and in x.

Stability Let $F(u) = cu$, $c = const$ and $\sigma = c\frac{\Delta t}{\Delta x}$. The FDE becomes

$$u_i^{n+1} = \frac{1}{2}\left[u_i^n + u_i^n - \sigma(u_{i+1}^n - u_i^n) - \sigma\left(u_i^n - \sigma(u_{i+1}^n - u_i^n)\right.\right.$$

$$\left.\left. - \left(u_{i-1}^n - \sigma(u_i^n - u_{i-1}^n)\right)\right)\right]$$

$$u_i^{n+1} = u_i^n - \frac{\sigma}{2}(u_{i+1}^n - u_{i-1}^n) + \frac{\sigma^2}{2}(u_{i+1}^n - 2u_i^n + u_{i-1}^n).$$

Carrying the complex mode $u_i^n = g^n e^{iia}$ in the equation gives

$$g = 1 - \sigma^2(1 - \cos\alpha) - i\sin\alpha.$$

Taking the modulus of g, after some algebra, yields

$$|g|^2 = 1 - 2(1 - \cos\alpha)\sigma^2(1 - \sigma^2).$$

The stability condition $|g|^2 \leq 1$ is satisfied when $\sigma^2 \leq 1$. This is the CFL condition. In term of the time step, the condition becomes:

$$\Delta t \leq \frac{\Delta x}{|c|}.$$

B.7.2 Exact Solution to Burgers' Equation (Inviscid)

General Solution Assume that $u(x,t) = G(x - tu(x,t))$. Taking partial derivatives in t and x gives

$$\begin{cases} \dfrac{\partial u}{\partial t} = \left(-u - t\dfrac{\partial u}{\partial t}\right)G' \\[2mm] \dfrac{\partial u}{\partial x} = \left(1 - t\dfrac{\partial u}{\partial x}\right)G' \end{cases}.$$

Then

$$\frac{\partial u}{\partial t} + u\frac{\partial u}{\partial x} = \left(-u - t\frac{\partial u}{\partial t} + u - tu\frac{\partial u}{\partial x}\right)G' = -t\left(\frac{\partial u}{\partial t} + u\frac{\partial u}{\partial x}\right)G'.$$

This can be rewritten as

$$(1 + tG')\left(\frac{\partial u}{\partial t} + u\frac{\partial u}{\partial x}\right) = 0.$$

Since, for arbitrary G, the first bracket is different from zero, it implies that Burgers' equation is satisfied.

Initial Value Problem The initial condition, applied to the unknown function G, gives

$$u(x,0) = G(x) = \begin{cases} 2\sqrt{-x}, & x \leq 0 \\ 0, & x \geq 0 \end{cases}.$$

$G(\xi)$ is determined for all values of its argument.

The characteristic lines for Burgers' equation are straight lines. Let x_0 be the intersection point of the characteristic with the x-axis. The equation of the characteristic is

$$\begin{cases} \phi(x,t) = x - x_0 - 2\sqrt{-x_0}t = 0, & x_0 \leq 0 \\ \phi(x,t) = x - x_0 = 0, & x_0 \geq 0 \end{cases}.$$

Two characteristics with different values of x_0 intersect. The solution is multi-valued. In order to make the solution single-valued, it is necessary to introduce a shock. See Fig. B.16.

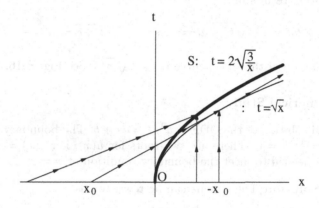

Fig. B.16. Characteristic lines, envelop and shock curve

Shock Curve The characteristics corresponding to $x_0 \geq 0$ do not admit an envelop, as they are parallel to the t-axis. For $x_0 \leq 0$, the envelop corresponds to the solution of the system

$$\begin{cases} x - x_0 - 2\sqrt{-x_0}t = 0 \\ -1 + \dfrac{1}{\sqrt{-x_0}}t = 0 \end{cases}.$$

A parametric representation of the envelop is given by

$$\begin{cases} t^* = \sqrt{-x_0} \\ x^* = -x_0 \end{cases}.$$

Upon elimination of x_0 the equation of the envelop becomes $t^* = \sqrt{x^*}$. See Fig. B.16. The envelop starts at $t = 0$, hence the shock will also appear at $t = 0$.

The shock condition states that $(dx/dt)_S = (u_1 + u_2)/2$. On the envelop, the slope is $(dx/dt)_E = 2\sqrt{x^*}$, the left state has velocity $u_1 = 2\sqrt{x^*}$ and the right side has velocity $u_2 = 0$. The jump condition is not satisfied.

At a point (x, t) on the shock, the left state must satisfy $u_1 = 2\sqrt{-(x - u_1 t)}$ since $\xi = x - u_1 t = \text{const}$ on the characteristic. Squaring this and solving for u_1 gives $u_1 = 2\left(t \pm \sqrt{t^2 - x}\right)$, where the plus sign root is the relevant root since for $t = 0$, $x = x_0$ the velocity is $u_1 = 2\sqrt{-x_0}$.

The differential equation for the shock is

$$\left(\frac{dx}{dt}\right)_S = t + \sqrt{t^2 - x}.$$

Assume that the shock curve is of the form $x = at^2$, $a > 0$, plugging this in the equation results in

$$2at = t + \sqrt{t^2 - at^2} \Rightarrow 2a = 1 + \sqrt{1 - a} \Rightarrow a = \frac{3}{4}.$$

The equation of the shock curve is $t = 2\sqrt{x/3}$. See Fig. B.16.

B.7.3 Numerical Study

The general solution of the ODE is $u^2/2 = ax + b$. The boundary conditions impose $a = \frac{1}{2}$, $b = 0$. There are two roots for $u(x)$, i.e. $u(x) = \pm\sqrt{x}$. The plus sign is needed to meet the boundary condition at $x = 1$.

Discrete Solution The scheme can be rewritten as

$$\frac{1}{2}\frac{u_{i+1}^2 - 2u_i^2 + u_{i-1}^2}{\Delta x^2} = \frac{1}{2}\frac{x_{i+1} - 2x_i + x_{i-1}}{\Delta x^2} = 0,$$

which proves that the exact solution satisfies the FDEs.

Iterative Method The computational molecule is shown in Fig. B.17.

Taylor expansion of the FD quotients, where $u^{n+1} - u^n = \Delta t(\partial u^n/\partial t) + ...$, gives

$$\frac{\Delta t}{\Delta x}\frac{\partial^2 u_i^n}{\partial t \partial x} + O\left(\frac{\Delta t^2}{\Delta x}, \Delta t\right) + \frac{\partial^2}{\partial x^2}\left(\frac{u^2}{2}\right)_i^n + O(\Delta x^2) = 0.$$

Choosing $\Delta t = \Delta x$ produces a meaningful evolution PDE

$$\frac{\partial^2 u}{\partial t \partial x} + \frac{\partial^2}{\partial x^2}\left(\frac{u^2}{2}\right) = 0.$$

To study the type of this PDE we introduce $v = \partial u/\partial x$, and transform the PDE into an equivalent first-order system

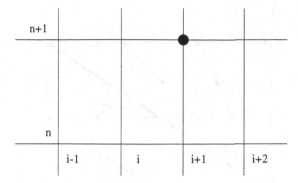

Fig. B.17. Computational molecule

$$\begin{cases} \dfrac{\partial u}{\partial x} = v \\ \dfrac{\partial v}{\partial t} + u\dfrac{\partial v}{\partial x} = -v^2 \end{cases}.$$

The characteristic matrix is obtained from the matrix form

$$\begin{bmatrix} 0 & 0 \\ 0 & 1 \end{bmatrix}\cdot\frac{\partial}{\partial t}\begin{bmatrix} u \\ v \end{bmatrix} + \begin{bmatrix} 1 & 0 \\ 0 & u \end{bmatrix}\cdot\frac{\partial}{\partial x}\begin{bmatrix} u \\ v \end{bmatrix} = \begin{bmatrix} v \\ -v^2 \end{bmatrix},$$

as

$$A = \begin{bmatrix} \dfrac{\partial \phi}{\partial x} & 0 \\ 0 & \dfrac{\partial \phi}{\partial t} + u\dfrac{\partial \phi}{\partial x} \end{bmatrix}.$$

The characteristic form is $Q = (\partial\phi/\partial x)\,((\partial\phi/\partial t) + u(\partial\phi/\partial x))$. There are two distinct roots. The system and the equation are totally hyperbolic.

The characteristic curves have slopes

$$\begin{cases} C^{\infty} : \left(\dfrac{dx}{dt}\right)_{C^{\infty}} = \infty \\ C^{+} : \left(\dfrac{dx}{dt}\right)_{C^{+}} = u > 0 \end{cases}.$$

The characteristic curves are sketched in Fig. B.18, with the arrows indicating the direction of propagation of the perturbations, as implied by the initial and boundary conditions. In particular, in the sub-region 1 bounded by the x-axis, the characteristic $x = t$ and the boundary $x = 1$, the solution $u = 1$ holds.

Solution to the Evolution Equation Integrating the second-order PDE in x gives $(\partial u/\partial t)+(\partial/\partial x)\left(u^2/2\right) = F(t)$, where $F(t)$ is an arbitrary function of t. In sub-region 1, $F = 0$.

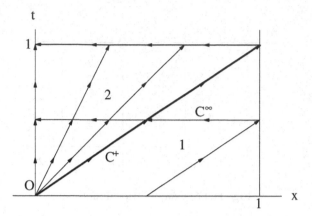

Fig. B.18. Characteristic lines in sub-regions 1 and 2

In sub-region 2, an expansion fan is the only solution possible that will connect the boundary condition at $x = 0$ to the characteristic $x = t$. $F = 0$ also in that sub-region. The solution for the expansion fan is the self-similar solution

$$u(x,t) = \frac{x}{t}.$$

The characteristics in sub-region 2 are shown in Fig. B.18.

Numerically, the solution will behave as the analytical solution above, until $t = 1$, when the characteristic $x = t$ hits the boundary. For $t > 1$ the characteristics are no longer straight lines, because $F \neq 0$. The solution continues to evolve until steady-state is reached. At steady-state $F = \frac{1}{2}$. The C^+ characteristics are now given by

$$\left(\frac{dx}{dt}\right)_{C^+} = \sqrt{x} \Rightarrow \phi(x,t) = t - 2\sqrt{x} = \text{const.}$$

They are parabolae with vertical tangent on the t-axis.

B.8 Solution to Problem 8

B.8.1 Analytic Study

System Type In matrix form the system of first-order PDEs reads

$$\begin{bmatrix} 1 & 0 \\ 0 & 1 \end{bmatrix} \cdot \frac{\partial}{\partial t} \begin{bmatrix} u \\ v \end{bmatrix} + \begin{bmatrix} 1 & 0 \\ 0 & -1 \end{bmatrix} \cdot \frac{\partial}{\partial x} \begin{bmatrix} u \\ v \end{bmatrix} + \begin{bmatrix} 0 & 1 \\ 1 & 0 \end{bmatrix} \cdot \frac{\partial}{\partial y} \begin{bmatrix} u \\ v \end{bmatrix} = \begin{bmatrix} 0 \\ 0 \end{bmatrix}.$$

The characteristic matrix follows as

$$A = \begin{bmatrix} \dfrac{\partial \phi}{\partial t} + \dfrac{\partial \phi}{\partial x} & \dfrac{\partial \phi}{\partial y} \\[3mm] \dfrac{\partial \phi}{\partial y} & \dfrac{\partial \phi}{\partial t} - \dfrac{\partial \phi}{\partial x} \end{bmatrix}.$$

The characteristic form is

$$Q = \det(A) = \left(\frac{\partial \phi}{\partial t} \right)^2 - \left(\left(\frac{\partial \phi}{\partial x} \right)^2 + \left(\frac{\partial \phi}{\partial y} \right)^2 \right).$$

There are always two real and distinct roots to $Q = 0$:

$$\frac{\partial \phi}{\partial t} = \pm \sqrt{\left(\frac{\partial \phi}{\partial x} \right)^2 + \left(\frac{\partial \phi}{\partial y} \right)^2}.$$

The system is totally hyperbolic.

Without restriction one can choose $\partial \phi / \partial x = \cos \theta$, $\partial \phi / \partial y = \sin \theta$, so that the roots are simply $\partial \phi / \partial t = \pm 1$.

The characteristic surfaces are planes perpendicular to the vector $\nabla \phi$. For a given value of θ there are two such planes passing through the origin:

$$\begin{cases} C^+ : & x \cos \theta + y \sin \theta + t = 0 \\ C^- : & x \cos \theta + y \sin \theta - t = 0 \end{cases}, \quad -\frac{\pi}{2} \le \theta \le \frac{\pi}{2}.$$

The planes make a 45° angle with the t-axis. They envelop a cone of half-angle 45°.

Compatibility Relations Replacing in $\det(A)$ one column by the equations r_1, r_2 gives

$$\left(\frac{\partial \phi}{\partial t} - \frac{\partial \phi}{\partial x} \right) r_1 - \frac{\partial \phi}{\partial y} r_2 = 0,$$

where $\nabla \phi$ corresponds to one of the roots, i.e.

$$\begin{cases} CR^+ : & (1 - \cos \theta)\, r_1 - \sin \theta \, r_2 = 0 \\ CR^- : & (1 + \cos \theta)\, r_1 + \sin \theta \, r_2 = 0 \end{cases}.$$

This is a homogeneous system for r_1, r_2. It is equivalent to the original system if the two relations are independent, that is if the determinant is different from zero. But

$$\begin{vmatrix} 1 - \cos \theta & -\sin \theta \\ 1 + \cos \theta & \sin \theta \end{vmatrix} = 2 \sin \theta,$$

which is zero for $\theta = 0$. If we introduce the half-angle $\frac{\theta}{2}$, the compatibility relations simplify to

$$\begin{cases} CR^+ : & \sin \frac{\theta}{2}\, r_1 - \cos \frac{\theta}{2}\, r_2 = 0 \\ CR^- : & \cos \frac{\theta}{2}\, r_1 + \sin \frac{\theta}{2}\, r_2 = 0 \end{cases},$$

which are independent.

Jump Conditions The jump conditions can be written as

$$\begin{cases} (n_t + n_x) \langle u \rangle + n_y \langle v \rangle = 0 \\ n_y \langle u \rangle + (n_t - n_x) \langle v \rangle = 0 \end{cases}.$$

This is a homogeneous linear system for $\langle u \rangle$, $\langle v \rangle$. If the jumps are not both zero, then the determinant of the system must be zero. It is the same determinant as that of the characteristic form for the components of the normal vector. Hence the discontinuities can only occur along characteristic planes, which is expected for a linear system.

Exact Solutions to Two Initial Value Problems i) The first initial value problem corresponds to 1-D flow:

$$u(x, y, 0) = v(x, y, 0) = 0, \ x < 0; \ u(x, y, 0) = v(x, y, 0) = 1, \ x > 0; \ \forall y.$$

The solution is independent of y. One can use the compatibility relations with $\theta = 0$, i.e.

$$\begin{cases} CR^+ : r_2 = \dfrac{\partial v}{\partial t} - \dfrac{\partial v}{\partial x} = 0 \text{ on } C^+ \\ \\ CR^- : r_1 = \dfrac{\partial u}{\partial t} + \dfrac{\partial u}{\partial x} = 0 \text{ on } C^- \end{cases}.$$

These can be integrated to give

$$\begin{cases} v(x, t) = F(x + t) \ on \ \xi = x + t = \text{const} \\ u(x, t) = G(x - t) \ on \ \eta = x - t = \text{const} \end{cases}.$$

The solution is

$$\begin{cases} u = v = 0, \ \xi = x + t < 0 \\ u = v = 1, \ \eta = x - t > 0 \\ u = 0, \ v = 1, \ \xi = x + t > 0, \ \eta = x - t < 0 \end{cases}.$$

The solution is shown in Figure B.19.

The jump conditions are satisfied since across the $C^+ : \ \xi = 0$, $\langle u \rangle = 0$, and across the $C^- : \ \eta = 0$, $\langle v \rangle = 0$.

ii) The second problem corresponds to

$$u(x, y, 0) = v(x, y, 0) = 1, \ y < 0; \ u(x, y, 0) = v(x, y, 0) = 0, \ y > 0; \ \forall x.$$

The solution is independent of x. Choosing $\theta = \frac{\pi}{2}$ in the compatibility relation yields

$$\begin{cases} CR^+ : r_1 - r_2 = \dfrac{\partial}{\partial t}(u - v) - \dfrac{\partial}{\partial y}(u - v) = 0 \text{ on } C^+ \\ \\ CR^- : r_1 + r_2 = \dfrac{\partial}{\partial t}(u + v) + \dfrac{\partial}{\partial y}(u + v) = 0 \text{ on } C^- \end{cases}.$$

These can be integrated to give

$$\begin{cases} u(y,t) - v(y,t) = F(y+t) \text{ on } \xi = y+t = \text{const} \\ u(y,t) + v(y,t) = G(y-t) \text{ on } \eta = y-t = \text{const} \end{cases}.$$

The solution is

$$\begin{cases} u = v = 1, \ \xi = y+t < 0 \\ u = v = 0, \ \eta = y-t > 0 \\ u = v = 1, \ \xi = y+t > 0, \ \eta = y-t < 0 \end{cases}.$$

The jump conditions are satisfied since across the $C^+ : \xi = 0$, $\langle u \rangle + \langle v \rangle = 0$, and across the $C^- : \eta = 0$, $\langle u \rangle - \langle v \rangle = 0$.

The solution is shown in Fig. B.19.

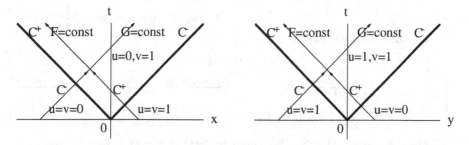

Fig. B.19. Solutions to the initial value problems

B.8.2 Numerical Study

Consistency, Accuracy The point chosen for the Taylor expansion of the first equation is the middle of the solid box $(i-\frac{1}{2}, j)$. However, the expansions are first made about the most convenient point as

$$\epsilon^n_{i-\frac{1}{2},j} =$$

$$\left(\frac{\partial u}{\partial t} + \frac{\Delta t}{2}\frac{\partial^2 u}{\partial t^2} + O(\Delta t^2)\right)^n_{i,j} + \left(\frac{\partial u}{\partial x} + O(\Delta x^2)\right)^n_{i-\frac{1}{2},j} + \left(\frac{\partial v}{\partial y} + O(\Delta y^2)\right)^{n+1}_{i-\frac{1}{2},j}.$$

Shifting the first and last term gives

$$\epsilon^n_{i-\frac{1}{2},j} =$$

$$\left(\frac{\partial u}{\partial t} + \frac{\partial u}{\partial x} + \frac{\partial v}{\partial y}\right)^n_{i-\frac{1}{2},j} + \left(\frac{\Delta t}{2}\frac{\partial^2 u}{\partial t^2} + \Delta t\frac{\partial^2 v}{\partial t\partial x} + \frac{\Delta x}{2}\frac{\partial^2 u}{\partial t\partial x} + O(\Delta x^2)\right)^n_{i-\frac{1}{2},j}$$

$$+ O\left((\Delta t + \Delta x)^2, \Delta y^2\right) = O(\Delta t, \Delta x, \Delta y^2).$$

The scheme is consistent, first-order accurate in t and x, and second-order accurate in y.

At steady-state the scheme is second-order accurate in both x and y.

It is clear that similar results are obtained for the second equation. The expansion are to be carried out about point $(i, j - \frac{1}{2})$. The results are the same as above when exchanging the roles of x and y.

Stability Let $\sigma_x = \Delta t/\Delta x$, $\sigma_y = \Delta t/\Delta y$. The scheme can be rewritten as

$$H \begin{bmatrix} U^{n+1} \\ V^{n+1} \end{bmatrix} = A \begin{bmatrix} U^n \\ V^n \end{bmatrix},$$

where

$$H = \begin{bmatrix} 1 & -\sigma_y(1 - \cos\beta - \underline{i}\sin\beta) \\ \sigma_y(1 - \cos\beta + \underline{i}\sin\beta) & 1 \end{bmatrix}$$

$$A = \begin{bmatrix} 1 - \sigma_x(1 - \cos\alpha + \underline{i}\sin\alpha) & 0 \\ 0 & 1 - \sigma_x(1 - \cos\alpha - \underline{i}\sin\alpha) \end{bmatrix}.$$

The amplification matrix is:

$$G = H^{-1}A =$$

$$\begin{bmatrix} \dfrac{1 - \sigma_x(1 - \cos\alpha + \underline{i}\sin\alpha)}{1 + 2\sigma_y^2(1 - \cos\beta)} & \dfrac{(1 - \sigma_x(1 - \cos\alpha - \underline{i}\sin\alpha))\sigma_y(1 - \cos\beta - \underline{i}\sin\beta)}{1 + 2\sigma_y^2(1 - \cos\beta)} \\ -\dfrac{(1 - \sigma_x(1 - \cos\alpha + \underline{i}\sin\alpha))\sigma_y(1 - \cos\beta + \underline{i}\sin\beta)}{1 + 2\sigma_y^2(1 - \cos\beta)} & \dfrac{1 - \sigma_x(1 - \cos\alpha - \underline{i}\sin\alpha)}{1 + 2\sigma_y^2(1 - \cos\beta)} \end{bmatrix}$$

The eigenvalues of G can be found by letting $\Lambda = (1 + 2\sigma_y^2(1 - \cos\beta))\lambda$. One finds Λ to be:

$$\Lambda = 1 - \sigma_x(1 - \cos\alpha)$$

$$\pm\sqrt{(\sigma_x\sin\alpha)^2 + ((1 - \sigma_x(1 - \cos\alpha))^2 + (\sigma_x\sin\alpha)^2)\sigma_y^2((1 - \cos\beta)^2 + \sin^2\beta)}.$$

In terms of λ one finds

$$|\lambda|^2 = \frac{1 - 2\sigma_x(1 - \sigma_x)(1 - \cos\alpha)}{1 + 2\sigma_y^2(1 - \cos\beta)}.$$

The stability condition is $0 \le \sigma_x \le 1$, independently of σ_y. This is a CFL condition on $\Delta t/\Delta x$. This result is understandable since the scheme is explicit in its treatment of the x-derivatives, but implicit in its treatment of the y-derivatives.

B.9 Solution to Problem 9

B.9.1 Analytical Study

The TSD equation reads

$$-(\gamma + 1)\frac{\partial\varphi}{\partial x}\frac{\partial^2\varphi}{\partial x^2} + \frac{\partial^2\varphi}{\partial y^2} = 0. \tag{B.3}$$

Transfer to the Hodograph The following mapping relations are easily derived:

$$\begin{cases} dx = \dfrac{\partial x}{\partial u}du + \dfrac{\partial x}{\partial v}dv \\[3mm] dy = \dfrac{\partial y}{\partial u}du + \dfrac{\partial y}{\partial v}dv \end{cases}.$$

These equations can be solved for du, dv as

$$\begin{cases} du = \dfrac{1}{J}\left(\dfrac{\partial y}{\partial v}dx - \dfrac{\partial x}{\partial v}dy \right) \\[3mm] dv = \dfrac{1}{J}\left(-\dfrac{\partial y}{\partial u}dx + \dfrac{\partial x}{\partial u}dy \right) \end{cases},$$

where $J = (\partial x/\partial u)(\partial y/\partial v) - (\partial y/\partial u)(\partial x/\partial v)$ is the Jacobian of the transformation.

From the above relations one infers that

$$\frac{\partial u}{\partial x} = \frac{1}{J}\frac{\partial y}{\partial v}, \quad \frac{\partial v}{\partial y} = \frac{1}{J}\frac{\partial x}{\partial u}.$$

Equation (B.3) can be written as a first-order PDE in (u, v)

$$-(\gamma + 1)u\frac{\partial u}{\partial x} + \frac{\partial v}{\partial y} = 0.$$

After multiplication by the Jacobian J, the governing equation becomes

$$\frac{\partial x}{\partial u} - (\gamma + 1)u\frac{\partial y}{\partial v} = 0. \tag{B.4}$$

Let $\Phi(u, v) = xu + yv - \varphi(x, y)$ be the Legendre potential. The partial derivatives of the Legendre potential are

$$\begin{cases} \dfrac{\partial \Phi}{\partial u} = x + u\dfrac{\partial x}{\partial u} + v\dfrac{\partial y}{\partial u} - \dfrac{\partial \varphi}{\partial x}\dfrac{\partial x}{\partial u} - \dfrac{\partial \varphi}{\partial y}\dfrac{\partial y}{\partial u} = x \\[3mm] \dfrac{\partial \Phi}{\partial v} = u\dfrac{\partial x}{\partial v} + y + v\dfrac{\partial y}{\partial v} - \dfrac{\partial \varphi}{\partial x}\dfrac{\partial x}{\partial v} - \dfrac{\partial \varphi}{\partial y}\dfrac{\partial y}{\partial v} = y \end{cases}.$$

Taking cross-derivatives of the Legendre potential gives a second first-order PDE

$$\frac{\partial y}{\partial u} - \frac{\partial x}{\partial v} = 0.$$

Finally, inserting x and y in (B.4) in terms of Φ gives the second-order potential equation

$$\frac{\partial^2 \Phi}{\partial u^2} - (\gamma + 1)u\frac{\partial^2 \Phi}{\partial v^2} = 0.$$

This is Tricomi's equation.

Equation Type We put the first-order system in matrix form:

$$\begin{bmatrix} 1 & 0 \\ 0 & 1 \end{bmatrix} \cdot \frac{\partial}{\partial u} \begin{bmatrix} x \\ y \end{bmatrix} + \begin{bmatrix} 0 & -(\gamma+1)u \\ -1 & 0 \end{bmatrix} \cdot \frac{\partial}{\partial v} \begin{bmatrix} x \\ y \end{bmatrix} = 0.$$

The characteristic matrix is

$$A = \begin{bmatrix} \dfrac{\partial \phi}{\partial u} & -(\gamma+1)u\dfrac{\partial \phi}{\partial v} \\[2mm] -\dfrac{\partial \phi}{\partial v} & \dfrac{\partial \phi}{\partial u} \end{bmatrix},$$

and the characteristic form $Q = (\partial \phi/\partial u)^2 - (\gamma+1)u\,(\partial \phi/\partial v)^2$. Note that $\phi(u, v)$ represents the equation of characteristic lines, if they exist. It is not to be confused with the potential $\Phi(u, v)$.

The two following cases arise:

i) $u < 0$, the characteristic form $Q = 0$ does not admit real roots. The equation is elliptic. This corresponds to subsonic flow,

ii) $u > 0$, two real and distinct roots of $Q = 0$ exist. The equation is totally hyperbolic. This corresponds to supersonic flow.

The sonic line is a transition line between elliptic and hyperbolic sub-domains. It corresponds to $u = 0$.

In the sub-domain $u > 0$, the slope of the characteristic curves is given by

$$\left(\frac{dv}{du}\right)_{C^\pm} = -\frac{\dfrac{\partial \phi}{\partial u}}{\dfrac{\partial \phi}{\partial v}} = \pm\sqrt{(\gamma+1)u}.$$

Integrating this ODE gives $v = \pm\frac{2\sqrt{\gamma+1}}{3}u^{\frac{3}{2}} + \text{const}$.

The characteristic curves are semi-cubics. See Fig. B.20.

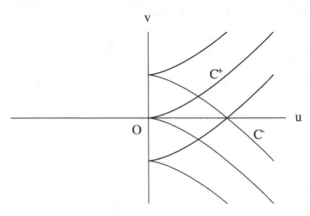

Fig. B.20. Characteristic lines of Tricomi's equation

B.9.2 Numerical Study

Consistency, Accuracy We consider each scheme in turn.

i) For this scheme the TE will be evaluated at point $(i-1, j)$, the scheme being centered there. We find

$$\epsilon_{i-1,j} = \frac{\partial^2 \Phi_{i-1,j}}{\partial u^2} + O(\Delta u^2) - (\gamma+1)u_{i-1}\frac{\partial^2 \Phi_{i-1,j}}{\partial v^2} + O(\Delta v^2) = O(\Delta u^2, \Delta v^2).$$

This scheme is consistent and second-order accurate in u and v.

ii) Choosing the same point we find

$$\epsilon_{i-1,j} = \frac{\partial^2 \Phi_{i-1,j}}{\partial u^2} + O(\Delta u^2) - (\gamma+1)(u_{i-1}+\Delta u)\frac{\partial^2 \Phi_{i-1,j}}{\partial v^2} + O(\Delta v^2)$$

$$= O(\Delta u, \Delta v^2).$$

This scheme is first-order accurate in u and second-order accurate in v.

iii) Here it is convenient to choose to expand about point (i, j) to get

$$\epsilon_{i,j} = \frac{\partial^2 \Phi_{i,j}}{\partial u^2} + O(\Delta u^2) - (\gamma+1)u_i\frac{\partial^2 \Phi_{i,j}}{\partial v^2} + O(\Delta v^2) = O(\Delta u^2, \Delta v^2).$$

This scheme is second-order accurate in u and v. It is consistent.

Iterative Method Schemes i) and ii) are explicit marching schemes. This is appropriate for supersonic flow.

Scheme iii) is a point-wise over-relaxation scheme. The values of Φ are obtained by sweeping the points in the subsonic sub-domain, for example starting from $i = 2$ and solving for the points along the column from $j = 1$ to $jx - 1$.

Stability Let $\sigma = \sqrt{(\gamma+1)u_{i-1}}(\Delta u/\Delta v)$, and $\Phi_{i,j} = g^i e^{ij\beta}$. Porting these in scheme i) gives

$$g^2 - 2\left(1 - \sigma^2(1 - \cos\beta)\right)g + 1 = 0.$$

This is the same equation obtained for the linear wave equation. The same CFL condition is required for stability, i.e. $\sigma \le 1$. The most restrictive condition will be obtained for the largest value of u_{i-1}, i.e. at the right boundary of the domain.

We do not anticipate difficulties at the sonic line because the slope of the characteristic vanishes there, in contrast to the TSD equation, where the slope is infinite at the sonic line.

B.9.3 Implicit Scheme

A simple SLOR scheme is analogous to the implicit scheme for the wave equation (see Fig. B.21):

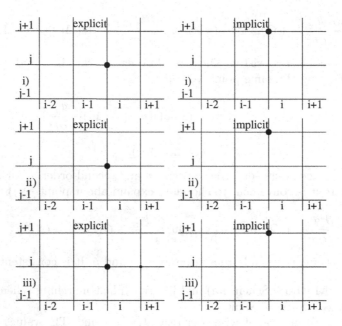

Fig. B.21. Sketch of computational molecule

i) $u_i > 0$ (supersonic point)

$$\frac{\Phi_{i,j}^{n+1} - 2\Phi_{i-1,j}^{n+1} + \Phi_{i-2,j}^{n+1}}{\Delta u^2} - (\gamma + 1)u_i \frac{\Phi_{i,j+1}^{n+1} - 2\Phi_{i,j}^{n+1} + \Phi_{i,j-1}^{n+1}}{\Delta v^2} = 0,$$

ii) $u_i \leq 0$ (subsonic point)

$$\frac{\Phi_{i+1,j}^{n} - 2\widetilde{\Phi}_{i,j} + \Phi_{i-1,j}^{n+1}}{\Delta u^2} - (\gamma + 1)u_i \frac{\Phi_{i,j+1}^{n+1} - 2\widetilde{\Phi}_{i,j} + \Phi_{i,j-1}^{n+1}}{\Delta v^2} = 0,$$

Note that the sonic point is no longer needed. It is treated as a regular supersonic point.

The matrix of the linear system is tridiagonal with diagonal dominance. The Thomas algorithm can be used to obtain the solution.

B.10 Solution to Problem 10

B.10.1 Analytic Study

Equation Type Let $u = \partial\varphi/\partial x$. The second-order PDE is transformed into two first-order PDEs that read

$$
\begin{cases}
\dfrac{\partial\varphi}{\partial x} = u \\[3mm]
\dfrac{\partial u}{\partial t} + \dfrac{\partial}{\partial x}\left(\dfrac{u^2}{2}\right) = 0
\end{cases}.
\tag{B.5}
$$

To study the type, assume that the solution is regular and expand the nonlinear term as

$$
\begin{cases}
\dfrac{\partial\varphi}{\partial x} = u \\[3mm]
\dfrac{\partial u}{\partial t} + u\dfrac{\partial u}{\partial x} = 0
\end{cases}.
$$

In matrix form we get

$$
\begin{bmatrix} 0 & 0 \\ 0 & 1 \end{bmatrix}\cdot\frac{\partial}{\partial t}\begin{bmatrix}\varphi \\ u\end{bmatrix}
+ \begin{bmatrix} 1 & 0 \\ 0 & u \end{bmatrix}\cdot\frac{\partial}{\partial x}\begin{bmatrix}\varphi \\ u\end{bmatrix}
= \begin{bmatrix} u \\ 0 \end{bmatrix}.
$$

The characteristic matrix can be constructed. Let $\phi(x,t)$ be the equation of a characteristic line, then

$$
A = \begin{bmatrix} \dfrac{\partial\phi}{\partial x} & 0 \\[3mm] 0 & \dfrac{\partial\phi}{\partial t} + u\dfrac{\partial\phi}{\partial x} \end{bmatrix},
$$

and the characteristic form is $Q = \frac{\partial\phi}{\partial x}\left(\frac{\partial\phi}{\partial t} + u\frac{\partial\phi}{\partial x}\right)$. There are two distinct roots to $Q = 0$. The system is totally hyperbolic.

The characteristic directions are given by

$$
\begin{cases}
C^\infty : \left(\dfrac{dx}{dt}\right)_{C^\infty} = \infty \Rightarrow \phi(x,t) = t = \text{const} \\[3mm]
C^u : \left(\dfrac{dx}{dt}\right)_{C^u} = u \Rightarrow \phi(x,t) = x - tu(x,t) = \text{const}
\end{cases}.
$$

We have used the fact that the characteristic lines of Burgers' equation (inviscid) are straight lines. The sketch of the characteristic lines for $u \geq 0$ and $u \leq 0$ is shown in Fig. B.22.

Jump Conditions The jump conditions are found from the conservative form of the system, equations (B.5). They read

$$\begin{cases} \langle\varphi\rangle\, n_x = 0 \\ \langle u\rangle\, n_t + \left\langle \dfrac{u^2}{2} \right\rangle n_x = 0 \end{cases}.$$

The first equation is trivial because the potential is continuous and $\langle\varphi\rangle = 0$. Hence, the jump conditions are those associated with Burgers' equation. They simplify to read

$$n_t + \frac{u_1 + u_2}{2} n_x = 0, \Rightarrow \left(\frac{dx}{dt}\right)_S = -\frac{n_t}{n_x} = \frac{u_1 + u_2}{2}.$$

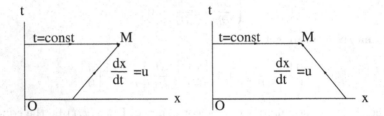

Fig. B.22. Characteristic lines of the PDE

Exact Solution for $u(x,t)$ In terms of $u(x,t)$ the initial conditions are

$$\begin{cases} u(x,0) = 1,\ 0 \le x \le \dfrac{1}{5} \\[2mm] u(x,0) = 0,\ \dfrac{1}{5} \le x \le \dfrac{1}{2} \\[2mm] u(x,0) = -\dfrac{1}{2},\ \dfrac{1}{2} \le x \le 1 \end{cases},$$

and the boundary conditions

$$\begin{cases} u(0,t) = 1 \\[2mm] u(1,t) = -\dfrac{1}{2} \end{cases}.$$

The initial and boundary conditions define the boundaries of subdomains as seen in Figure B.23.

At $x = \frac{1}{5}$ there is a discontinuity with decreasing velocity in the initial data, with constant states on either sides. Hence a shock S_1 originates there. Its slope is $\left(\frac{dx}{dt}\right)_{S_1} = \frac{1}{2}$.

At $x = \frac{1}{2}$ there is a discontinuity, also with decreasing velocity, and constant states on either sides. The shock S_2 has a slope $(dx/dt)_{S_2} = -\frac{1}{4}$.

S_1 and S_2 are boundaries between uniform flow regions. They meet at point $(\frac{2}{5}, \frac{2}{5})$. When the two shocks meet, a single, stronger shock is formed, S_3 with slope $\frac{1}{4}$. See Fig. B.23.

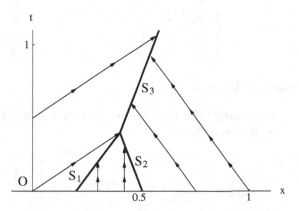

Fig. B.23. Shock lines and uniform flow regions

The solution, in the three regions, is given by

$$
\begin{cases}
u_1(x,t) = 1, & \begin{cases} x - \frac{1}{2}t \le \frac{1}{5},\ t \le \frac{2}{5} \\ x - \frac{1}{4}t \le \frac{3}{10},\ t \ge \frac{2}{5} \end{cases} \\[2em]
u_2(x,t) = 0,\ x - \frac{1}{2}t \ge \frac{1}{5},\ x + \frac{1}{4}t \le \frac{1}{2} \\[1em]
u_3(x,t) = -\frac{1}{2}, & \begin{cases} x + \frac{1}{4}t \ge \frac{1}{2},\ t \le \frac{2}{5} \\ x - \frac{1}{4}t \ge \frac{3}{10},\ t \ge \frac{2}{5} \end{cases}
\end{cases}
$$

Exact Solution for $\varphi(x,t)$ The solution $\varphi(x,t)$ is obtained by integration in x using the boundary condition at $x = 0$, i.e. $\varphi(x,t) = \int_0^x u(\xi,t)d\xi$.

One finds:

$$
\begin{cases}
\varphi_1(x,t) = x, & \begin{cases} x - \dfrac{1}{2}t \leq \dfrac{1}{5},\ t \leq \dfrac{2}{5} \\[2mm] x - \dfrac{1}{4}t \leq \dfrac{3}{10},\ t \geq \dfrac{2}{5} \end{cases} \\[6mm]
\varphi_2(x,t) = \dfrac{1}{5} + \dfrac{1}{2}t,\ x - \dfrac{1}{2}t \geq \dfrac{1}{5},\ x + \dfrac{1}{4}t \leq \dfrac{1}{2} \\[6mm]
\varphi_3(x,t) = \begin{cases} \dfrac{11}{20} + \dfrac{1}{8}t - \dfrac{1}{2}x,\ x + \dfrac{1}{4}t \geq \dfrac{1}{2},\ t \leq \dfrac{2}{5} \\[2mm] \dfrac{9}{20} + \dfrac{3}{8}t - \dfrac{1}{2}x,\ x - \dfrac{1}{4}t \geq \dfrac{3}{10},\ t \geq \dfrac{2}{5} \end{cases}
\end{cases}
$$

B.10.2 Numerical Scheme

Consistency, Accuracy The first scheme is best expanded about point $(i-1, n)$. Let's write the first term as

$$
\frac{\varphi_i^{n+1} - \varphi_{i-1}^{n+1} - \varphi_i^n + \varphi_{i-1}^n}{\Delta t \Delta x} = \frac{\dfrac{\varphi_i^{n+1} - \varphi_{i-1}^{n+1}}{\Delta x} - \dfrac{\varphi_i^n - \varphi_{i-1}^n}{\Delta x}}{\Delta t}
$$

$$
= \frac{1}{\Delta t}\left(\frac{\partial \varphi}{\partial x} + \frac{\Delta x}{2}\frac{\partial^2 \varphi}{\partial x^2} + O(\Delta x^2)\right)^{n+1}_{i-1} - \frac{1}{\Delta t}\left(\frac{\partial \varphi}{\partial x} + \frac{\Delta x}{2}\frac{\partial^2 \varphi}{\partial x^2} + O(\Delta x^2)\right)^n_{i-1}
$$

$$
= \frac{\partial^2 \varphi_{i-1}^n}{\partial t \partial x} + \frac{\Delta t}{2}\frac{\partial^3 \varphi_{i-1}^n}{\partial t^2 \partial x} + O(\Delta t^2) + \frac{\Delta x}{2}\frac{\partial^3 \varphi_{i-1}^n}{\partial t \partial x^2} + O(\Delta t \Delta x, \Delta x^2).
$$

The second FD quotient is centered at $(i-1, n)$ and is second-order accurate there. The TE becomes

i)

$$
\epsilon_{i-1}^n = \frac{\Delta t}{2}\frac{\partial^3 \varphi_{i-1}^n}{\partial t^2 \partial x} + \frac{\Delta x}{2}\frac{\partial^3 \varphi_{i-1}^n}{\partial t \partial x^2} + O(\Delta t + \Delta x)^2.
$$

The scheme is consistent and first-order accurate in t and x.

ii) This scheme is not consistent since the space derivative is doubled. This is a well known feature of the Murman (1973) scheme. It is conservative, which is more important as it will compute the correct shock speed.

iv) Expanding about (i, n) which is the centered point for the second FD quotient which is second-order accurate there, and using the symmetries of the first term, one finds

$$
\epsilon_i^n = -\frac{\Delta t}{2}\frac{\partial^3 \varphi_i^n}{\partial t^2 \partial x} - \frac{\Delta x}{2}\frac{\partial^3 \varphi_i^n}{\partial t \partial x^2} + O(\Delta t + \Delta x)^2.
$$

The scheme is first-order accurate in t and x.

At steady-state, schemes i) and iv) are second-order accurate in x.

Stability The stability of the shock point is studied. Let $\sigma^- = u_{i-1}(\Delta t/\Delta x)$, $\sigma^+ = -u_i(\Delta t/\Delta x)$, $\varphi_i^n = g^n e^{ii\alpha}$. One obtains

$$g = \frac{(1 - \cos\alpha)(1 + 2\sigma^- \cos\alpha - 2\sigma^+) + i(1 - 2\sigma^-(1 - \cos\alpha))\sin\alpha}{1 + \cos\alpha + i\sin\alpha}.$$

Taking the modulus, after some algebra we arrive at

$$|g|^2 = 1 - 2\sigma^- - 2\sigma^+ + 2(\sigma^-)^2 + 2(\sigma^+)^2$$

$$+2\left(\sigma^- + \sigma^+ - 2\sigma^-\sigma^+ - (\sigma^-)^2 - (\sigma^+)^2\right)\cos\alpha + 4\sigma^-\sigma^+ \cos^2\alpha.$$

For stability we need $|g|^2 \leq 1$, hence the expression E must verify

$$E = -\sigma^- - \sigma^+ + (\sigma^-)^2 + (\sigma^+)^2 + \left(\sigma^- + \sigma^+ - 2\sigma^-\sigma^+ - (\sigma^-)^2 - (\sigma^+)^2\right)$$

$$\cdot \cos\alpha + 2\sigma^-\sigma^+ \cos^2\alpha \leq 0.$$

This is a quadratic function of $\cos\alpha$, $-1 \leq \cos\alpha \leq 1$. It suffices that $E(1) \leq 0$, $E(-1) \leq 0$ to satisfy the above condition. We find

$$\begin{cases} E(1) = 0 \\ E(-1) = -2(\sigma^- + \sigma^+ - 2\sigma^-\sigma^+ - (\sigma^-)^2 - (\sigma^+)^2) \\ \quad\quad = -2(\sigma^- + \sigma^+)(1 - (\sigma^- + \sigma^+)) \end{cases}.$$

The first condition is satisfied. The second condition requires $\sigma^- + \sigma^+ \leq 1$. This is a CFL condition for the time step.

The result can now be found for the various schemes:

ii) shock point

$$\Delta t \leq \frac{\Delta x}{u_{i-1} - u_i},$$

i) supersonic point

$$\Delta t \leq \frac{\Delta x}{u_{i-1}},$$

iv) subsonic point

$$\Delta t \leq -\frac{\Delta x}{u_i}.$$

B.10.3 Implicit Scheme

The scheme can be made implicit with a fixed-point approach, by shifting the space derivative from level n to level $n + 1$. The computational molecule is shown in Fig. B.24.

Consistency, Accuracy Due to the symmetries of the computational molecule for the implicit scheme in reference to the explicit scheme, we expect to have the same results, i.e. schemes i) and iv) are consistent and first-order accurate in x and t, scheme ii) is not consistent.

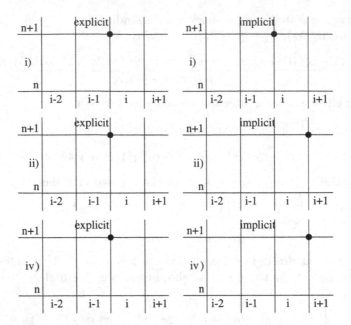

Fig. B.24. Sketch of computational molecule

At steady-state i) and iv) are second-order accurate in x.

Matrix Structure The matrix has at most four diagonals. It can be solved using a specialized Gaussian elimination technique for penta-diagonal matrices, which is an extension of the Thomas algorithm.

B.11 Solution to Problem 11

Consider the following PDE, IC and BCs for $u(x,t)$:

$$\frac{\partial u}{\partial t} + \frac{\partial u}{\partial x} = (1-u)\frac{g'}{g},$$
$$u(x,0) = 0,\ 0 \le x \le 1,$$
$$u(0,t) = 0,\ t \ge 0.$$

$g(x)$ represents the area of the nozzle diffuser, where $g(0) = 1$ and $g(x) \ge 1$, $0 \le x \le 1$.

B.11.1 Analysis of the Problem

Equation Type The PDE is linear. As a single first-order PDE it is hyperbolic. Indeed, the characteristic matrix is $A = (\frac{\partial \phi}{\partial t} + \frac{\partial \phi}{\partial x})$, a (1×1) matrix, and the characteristic form admits a real root $\frac{\partial \phi}{\partial t} = -\frac{\partial \phi}{\partial x}$. The type does not depend on $g(x)$.

Characteristic Lines The characteristic line has slope $\left(\frac{dx}{dt}\right)_C = -\frac{\partial \phi}{\partial t} / \frac{\partial \phi}{\partial x} = 1$. The characteristics are the straight lines of equation $\xi = x - t = \text{const}$.

Compatibility Relation The compatibility relation is the equation itself. If $u_c = u[x, t_c(x)]$ then

$$\frac{du_c}{dx} = \frac{\partial u}{\partial x} + \frac{\partial u}{\partial t}\left(\frac{dt_c}{dx}\right) = \frac{\partial u}{\partial x} + \frac{\partial u}{\partial t} \Rightarrow \frac{du_c}{dx} = \frac{\partial u}{\partial t} + \frac{\partial u}{\partial x} = (1 - u_c)\frac{g'}{g}.$$

This is an ODE. The variables can be separated to give

$$\frac{du_c}{1 - u_c} = \frac{dg}{g} \Rightarrow -\ln|1 - u_c| = \ln|g| + \text{const} \Rightarrow 1 - u_c = \frac{C}{g} \Rightarrow u_c = 1 - \frac{C}{g}.$$

The constant C depends on the initial or boundary condition.

Exact Solution The characteristic $\xi = x - t = 0$ divides the domain into two subdomains 1 ($\xi \geq 0$) and 2 ($\xi \leq 0$). See Fig. B.25.

In subdomain 1 the characteristics originate from the x-axis where the initial condition holds. Hence

$$u_c(x) = u(x, 0) = 0 = 1 - \frac{C}{g(x)}, 0 \leq x \leq 1 \Rightarrow C(\xi) = g(\xi) \Rightarrow u_c(\xi) = 1 - \frac{g(\xi)}{g(x)}.$$

In subdomain 1 the solution is

$$u(x, t) = 1 - \frac{g(x - t)}{g(x)}, \ 0 \leq x - t \leq 1.$$

In subdomain 2 the characteristics originate from the t-axis where the boundary condition holds. Hence

$$u_c(t) = u(0, t) = 0 = 1 - \frac{C}{g(0)} = 1 - C, t \geq 0 \Rightarrow C = 1 \Rightarrow u_c(\xi) = 1 - \frac{1}{g(x)}.$$

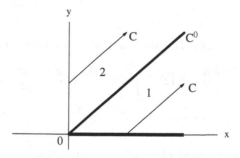

Fig. B.25. Subdomains and characteristic lines

In subdomain 2 the exact solution is

$$u(x,t) = 1 - \frac{1}{g(x)}, \quad x - t \leq 0.$$

Jump Conditions The jump conditions are simply $\langle u \rangle n_t + \langle u \rangle n_x = 0$, where $\vec{n} = (n_x, n_t)$ is the unit normal vector to the jump line. The slope of the jump line L is given by

$$\left(\frac{dx}{dt}\right)_L = -\frac{n_t}{n_x} = 1 = \left(\frac{dx}{dt}\right)_C.$$

The jump takes place along a characteristic.

B.11.2 Explicit Scheme

Consider the following scheme

$$\frac{u_i^{n+1} - u_i^n}{\Delta t} + \frac{u_i^n - u_{i-1}^n}{\Delta x} - (1 - u_{i-1}^n)\frac{g_i - g_{i-1}}{\Delta x \, g_i} = 0.$$

The sketch of the computational molecule is shown in Fig. B.26.

The speed of propagation of the perturbations on the characteristic C is $(dx/dt)_C = 1$. On the numerical characteristic it is $(dx/dt)_N = (\Delta x/\Delta t)$.

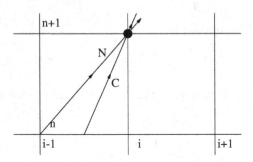

Fig. B.26. Computational molecule for the explicit scheme

Consistency, Accuracy Expand the truncation error about (i, n):

$$\epsilon_i^n = \left\{\frac{\partial u}{\partial t} + \frac{\Delta t}{2}\frac{\partial^2 u}{\partial t^2} + O(\Delta t^2)\right\}_i^n + \left\{\frac{\partial u}{\partial x} - \frac{\Delta x}{2}\frac{\partial^2 u}{\partial x^2} + O(\Delta x^2)\right\}_i^n$$
$$-\left\{1 - \left(u - \Delta x\frac{\partial u}{\partial x} + O(\Delta x^2)\right)\right\}_i^n \frac{1}{g_i}\left\{g' - \frac{\Delta x}{2}g'' + O(\Delta x^2)\right\}_i^n.$$

After reordering of the product of the Taylor expansions in the last term, the truncation error reads

$$\epsilon_i^n = \left\{ \frac{\partial u}{\partial t} + \frac{\partial u}{\partial x} - (1-u)\frac{g'}{g} \right\}_i^n + \frac{\Delta t}{2}\left(\frac{\partial^2 u}{\partial t^2}\right)_i^n - \frac{\Delta x}{2}\left(\frac{\partial^2 u}{\partial x^2}\right)_i^n +$$

$$\frac{\Delta x}{2}(1-u_i^n)\frac{g_i''}{g_i} - \Delta x \left(\frac{\partial u}{\partial x}\right)_i^n \frac{g_i'}{g_i} + O(\Delta t^2, \Delta x^2).$$

The terms in the first bracket cancel out, since u is the exact solution and the error is of the form $O(\Delta t, \Delta x)$ indicating that the scheme is consistent and first-order accurate in x and in t.

At steady-state, the error reduces to

$$\epsilon_i = -\frac{\Delta x}{2}\left\{ u'' - (1-u)\frac{g''}{g} + 2u'\frac{g'}{g} \right\}_i + O(\Delta x^2).$$

However, the exact solution verifies

$$u' = (1-u)\frac{g'}{g} \quad \text{and} \quad u'' = -u'\frac{g'}{g} + (1-u)\frac{g''}{g} - (1-u)\frac{g'^2}{g^2}.$$

Upon substitution, one finds that the terms in the bracket cancel out and the error is $\epsilon_i = O(\Delta x^2)$ indicating that the scheme is of second- or higher-order accuracy.

In order to conclude, one substitutes the exact steady-state solution from part 1.4, i.e. $u_i = 1 - \frac{1}{g_i}$ into the finite difference scheme to get

$$\epsilon = \frac{(1-\frac{1}{g_i}) - (1-\frac{1}{g_{i-1}})}{\Delta x} - \left(1 - \left(1 - \frac{1}{g_{i-1}}\right)\right)\frac{g_i - g_{i-1}}{\Delta x\, g_i}$$

$$= \frac{1}{\Delta x}\left\{ \frac{1}{g_{i-1}} - \frac{1}{g_i} - \frac{g_i - g_{i-1}}{g_{i-1}g_i} \right\} \equiv 0.$$

The scheme computes the exact solution at all points, regardless of the mesh step.

Stability The Von Neumann analysis consists in substituting $u_i^n = U^n e^{ii\alpha}$ in the update form $u_i^{n+1} = u_i^n - \sigma(u_i^n - u_{i-1}^n)$ where the equation has been made homogeneous and $\sigma = \frac{\Delta t}{\Delta x} > 0$. Hence

$$U = 1 - \sigma(1 - e^{-i\alpha}) = 1 - \sigma(1 - \cos\alpha) - i\sigma\sin\alpha.$$

Taking the modulus of the amplification factor gives

$$|U|^2 = [1 - \sigma(1-\cos\alpha)]^2 + [\sigma\sin\alpha]^2 = 1 - 2(1-\cos\alpha)\sigma(1-\sigma).$$
$$|U|^2 \le 1 \iff \sigma \le 1.$$

The scheme will be stable when it satisfies the CFL (Courant-Friedrichs-Lewy) condition $\Delta t \le \Delta x$.

B.11.3 Implicit Scheme

The following implicit scheme is proposed:

$$\frac{u_i^{n+1} - u_i^n}{\Delta t} + \frac{u_i^{n+1} - u_{i-1}^{n+1}}{\Delta x} - (1 - u_{i-1}^{n+1})\frac{g_i - g_{i-1}}{\Delta x\, g_i} = 0.$$

The computational molecule is sketched in Fig. B.27. The speed of propagation of the perturbation on the C characteristic is $(\frac{dx}{dt})_C = 1$. The speed is infinite on the N characteristic, i.e. $(\frac{dx}{dt})_N = +\infty$.

Consistency, Accuracy The implicit scheme is a mirror image of the explicit scheme. The Taylor expansion about $(i, n+1)$ will produce a similar truncation error, except for some sign changes (one can replace Δt by $-\Delta t$ in the previous result).

During the transient, the scheme is first-order accurate in x and t.

At steady-state, the scheme is identical to the previous one: it is exact! $\epsilon \equiv 0$.

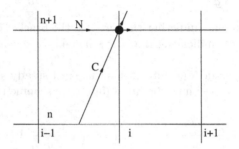

Fig. B.27. Computational molecule for the implicit scheme

Stability In update form, the equation reads $u_i^{n+1} + \sigma(u_i^{n+1} - u_{i-1}^{n+1}) = u_i^n$. The amplification factor satisfies $[1 + \sigma(1 - e^{-i\alpha})]U = 1$. Solving for U gives

$$U = \frac{1}{1 + \sigma(1 - \cos\alpha) + i\sigma\sin\alpha},$$

which has a modulus always less or equal to 1, regardless of σ and the time step. The implicit scheme is unconditionally stable.

Solution Procedure The matrix associated with this implicit scheme has a lower bi-diagonal structure. There is no need for a solver. It is solved by marching the scheme with increasing values of the index i.

The coefficients of the matrix and the right-hand-side are

$$p_i = -\frac{1}{\Delta x} + \frac{g_i - g_{i-1}}{\Delta x\, g_i}, \quad q_i = \frac{1}{\Delta t} + \frac{1}{\Delta x}, \quad r_i = 0,$$

$$s_i = -\frac{u_i^n - u_{i-1}^n}{\Delta x} + (1 - u_{i-1}^n)\frac{g_i - g_{i-1}}{\Delta x\, g_i}.$$

References

1. Kreyszig, E., *Advanced Engineering Mathematics*, 8th ed. New-York, Wiley (1998).
2. Courant, R. and Hilbert, D., *Methods of Mathematical Physics*, Vol. II, Partial Differential Equations, J. Wiley & Sons (1989).
3. Roache,P.J., *Computational Fluid Dynamics*, Hermosa Publishers, Albuquerque, NM (1972).
4. Cole, J.D., *Twenty Years of Transonic Flow*, Boeing Scientific Research Labs D1-82-0878 (1969).
5. Murman, E.M. and Cole, J.D., Calculation of Plane Steady Transonic Flows, *AIAA Journal*, **9**, No.1 (1971).
6. Murman, E.M., Analysis of Embedded Shock Waves Calculated by Relaxation Methods, *AIAA Journal*, **12**, No.5 (1974).
7. Chattot, J.J., A conservative Box-Scheme for the Euler Equations, *Int. J. Numer. Meth. Fluids* **31**, 149-158 (1999).
8. Roe, P.L., Approximate Riemann Solvers, Parameter vectors and Difference Schemes, *J. Comput. Phys.*, 43, No.2, 357-372 (1981).
9. Holt, M., *Numerical Methods in Fluid Dynamics*, Springer Series in Computational Physics, Springer-Verlag (1977).
10. Peyret, R. and Taylor, T.D., *Computational Methods for Fluid Flow*, Springer Series in Computational Physics, Springer-Verlag (1990).

Index

Scientific Computation